MOUNTAIN LIONS OF THE BLACK HILLS

Mountain Lions of the Black Hills

HISTORY AND ECOLOGY

JONATHAN A. JENKS

Johns Hopkins University Press
Baltimore

Johns Hopkins University Press
2715 North Charles Street
Baltimore, Maryland 21218-4363
www.press.jhu.edu

Library of Congress Cataloging-in-Publication Data
Names: Jenks, Jonathan Alden, 1954–, author.
Title: Mountain lions of the Black Hills : history and ecology / Jonathan A. Jenks.
Description: Baltimore : Johns Hopkins University Press, 2018. | Includes bibliographical
   references and index.
Identifiers: LCCN 2017016622| ISBN 9781421424422 (hardcover : alk. paper) |
   ISBN 9781421424439 (electronic) | ISBN 1421424428 (hardcover : alk. paper) |
   ISBN 1421424436 (electronic)
Subjects: LCSH: Puma—Black Hills Region (S.D. and Wyo.)
Classification: LCC QL737.C23 J46 2018 | DDC 599.75/24—dc23
   LC record available at https://lccn.loc.gov/2017016622

A catalog record for this book is available from the British Library.

Frontispiece: Photograph by Dan Thompson

*Special discounts are available for bulk purchases of this book. For more information, please contact
Special Sales at 410-516-6936 or specialsales@press.jhu.edu.*

Johns Hopkins University Press uses environmentally friendly book materials, including
recycled text paper that is composed of at least 30 percent post-consumer waste, whenever
possible.

CONTENTS

PREFACE

The chapters of this book cover differing but in some aspects overlapping objectives for the study of mountain lions occupying the Black Hills region. In chapters 1 and 2, I provide some history of the Black Hills region, beginning with the Custer reconnaissance in 1874, when the US Army surveyed the region with the aim, first, to document resources such as geology and vegetation and then, later, in part, to find potential food resources. In chapter 2 I present an overview of the ecology of the region, including typical plants and animals, both of which were originally diverse; later restoration efforts bolstered the elk and bighorn sheep populations currently inhabiting the region.

In chapter 3 the lead species, the mountain lion, is introduced, and I provide information to characterize the species, including how Black Hills lions are different from other populations. As readers will ascertain, lions, because of their vast distribution, are similar throughout North America. However, seasoned biologists take note of their most noteworthy or redeeming characteristics, especially those related to the best techniques for capturing them for research purposes and how they use their particular environment in their daily activities.

Chapters 4 through 7 are focused on the biology of the species. I start by providing information on population dynamics, such as population growth, survival, and reproduction, and also explain how the species changed as it recolonized the Black Hills. Chapter 5, which deals with disease ecology, has a link with chapter 4, on population dynamics, because there was a potential for diseases to arise as the mountain lion population became relatively dense in the region. Nutritional ecology, the topic of chapter 6, can be considered basic to the characterization of mountain lions over the entire recolonization period. The same is true of the genetics of mountain lions, detailed in chapter 7, because we collected blood and tissue samples from captured mountain lions throughout our studies as well as from numerous lion carcasses obtained from the South Dakota Department of Game, Fish and Parks.

In chapter 8 I discuss how mountain lions were perceived by South Dakotans, both early in our work and later, when the species became commonplace because of its abundance. The harvest of the species has elicited public comment, and recreationalists have viewed the predator negatively because they believed it caused a reduction in the harvest allotment of game animals, in particular, elk. Finally, the epilogue summarizes how the long-term nature of our work allowed us to better understand the inner and outer workings of this secretive large carnivore and how it recolonized a semi-isolated region of the country, the Black Hills.

ACKNOWLEDGMENTS

The long-term nature of the information in this book required support from numerous individuals and agencies. I am especially indebted to the administration and employees of the South Dakota Department of Game, Fish and Parks for funding projects on mountain lions, beginning in 1998 and extending through 2014. These projects (study numbers 7594, 75106, 7587, 7537, and 7545) were funded through the Federal Aid to Wildlife Restoration Act, administered through the South Dakota Department of Game, Fish and Parks. I thank Doug Hanson, George Vandel, Tony Leif, Emmitt Keyser, Tom Kirschenmann, Ron Fowler, Chad Switzer, Mike Kintigh, John Kanta, Steve Griffin, Jack Alexander, Ted Benzon, Andy Lindbloom, Chad Lehman, Blair Waite, and numerous others who helped with mountain lion captures and supported our work within the department. At times, the Rapid City Chapter of Safari Club International also provided support for our projects. Also, I am grateful to the many volunteers who helped with lion captures and the many landowners who provided access to their property to allow capturing mountain lions and locating or retrieving their carcasses.

Individuals associated with the US Department of Agriculture (USDA) Black Hills National Forest, the Wind Cave National Park, the Jewel Cave National Monument, the Mount Rushmore National Monument, the Black Elk Wilderness Area, the Wyoming Game and Fish Department, the North Dakota Game and Fish Department, the Montana Department of Parks and Wildlife, the Nebraska Department of Game and Parks, the Oklahoma Department of Wildlife Conservation, the Minnesota Department of Natural Resources, and the Manitoba Ministry of Natural Resources also provided information or support for our work at one time or another. These studies could not have been completed without the aid of the South Dakota Wing of the Civil Air Patrol and its many pilots, including Leo Becht, Rodney (Buck) Deweese, Gerald Kirk, and Mike Beason.

I also thank my graduate students, Dorothy (Fecske) Wells, resource biologist, Great Swamp National Wildlife Refuge; Dan Thompson, leader, Trophy Game Office, Wyoming Game and Fish Department; Brian Jansen, Arizona Game and Fish Department; Josh Smith, Savanah River Ecology Laboratory; and Beckie Juarez, Savannah River Ecology Laboratory, who completed graduate degrees working on mountain lions in the Black Hills. These individuals collected much of the information presented in the book, contributed to critical discussions on many aspects of these projects, and ensured the success of their projects. We learned much about the species in a relatively short period of time.

The staff of the Department of Natural Resource Management (formerly the Department of Wildlife and Fisheries Sciences), including department heads Chuck Scalet and Dave Willis, helped with administrative issues that always come up when fieldwork is conducted through university systems. Terri Symens, Diane Drake, Carol Doyle, and Dawn Van Ballegooyen (department staff) provided students with the information they needed to meet the expectations of the department and guidelines for budgetary information for their projects, thus allowing the students to focus on the collection of data. All projects (proposal numbers 99-A002, 99-A010, 02-A034, 05-E013, 07-A046, 09-019A, 09-079A, 11-002A, 11-078A, and 14-005A) were approved by the Institutional Animal Care and Use Committee at South Dakota State University.

I am indebted to Tom Kirschenmann, Dan Thompson, Larry Gigliotti, John Kanta, and two anonymous reviewers who provided critical comments on and suggestions for earlier drafts of the manuscript. I appreciate their suggestions, which helped to ensure that the information provided is correct. I am especially indebted to Dan Thompson, who went out of his way to obtain needed information and provide photos for the book. Dorothy (Fecske) Wells also helped in this regard. Our dear friends Kari and Jerry Olson also provided comments on an early draft of the manuscript. Eric Michel, Kyle Kaskie, and Virginia Coudron provided much-needed help with formatting figures for publication. I thank the staff at Johns Hopkins University Press, including Vince Burke, Tiffany Gasbarrini, and Mary Lou Kenney, who provided substantial support throughout the development, review, and production of the book. I am grateful to Lois Crum for her expert copyediting.

I also thank my mentors, David M. "Chip" Leslie Jr., R. Terry Bowyer, and Ron Barry, for their support, which has continued throughout my career. Finally, and most importantly, I thank my wife, Gail, for her support over the past 44 years, and of course my children and their spouses, Heather and Tad, Jonathan and Erin, and Abigail and Brandon, and my grandchildren, Andrew, Brynn, Bailey, Bode, and Boston.

MOUNTAIN LIONS OF THE BLACK HILLS

CHAPTER 1

Introduction

I was first introduced to the Black Hills in 1991 as a new assistant professor at South Dakota State University (SDSU). I had just earned a PhD from Oklahoma State University, where I had studied white-tailed deer nutrition on lands in southeastern Oklahoma and southwestern Arkansas owned and operated by the Weyerhaeuser Company. When I arrived at SDSU, I acquired three students who were working on deer and elk (*Cervus elaphus*) in the Black Hills. Two of them were in the midst of collecting data for their projects, one in the northern Black Hills and the other in Custer State Park. My department head, Chuck Scalet, strongly suggested that I travel to the Black Hills to meet the students and to get a feel for this new (to me) state of South Dakota, where I would be devoting much of my time to conducting research and teaching students about wildlife ecology and management.

My son, Jonathan, and I piled into our Mercury Lynx and headed west from Brookings, South Dakota, just a few weeks after arriving in the state. It was late July, and we traveled due west through the corn and soybean fields of eastern South Dakota to Pierre and then across the high plains of western South Dakota to Sturgis, which is on the outskirts of the northern Black Hills (fig. 1.1). After meeting with one of my new students, along with employees of South Dakota Game, Fish and Parks and Wyoming Game and Fish, we headed south through the timbered mountains (hills) to Mount Rushmore (a diversion from my objective but a must-see for both my son and me), and finally to Custer State Park, where I met my second student. My initial feeling for the Black Hills was essentially one of going home. I was raised on a dairy farm in eastern Massachusetts and had spent nine years in Maine obtaining bachelor's and master's degrees from Unity College and the University of Maine, respectively, studying wildlife ecology and management. My master's project involved accepting abandoned fawns to establish a captive herd of white-tailed deer (*Odocoileus*

FIGURE 1.1. View of the Black Hills from Route 212 looking to the southwest and not far from the site where scouts from Lt. Col. George Armstrong Custer's reconnaissance of the Black Hills first viewed the region in 1874. *Photo by Adam Kauth.*

virginianus) and evaluating their use of arboreal lichen (*Usnea* sp.) (Jenks and Leslie 1988, 1989). The Black Hills region was similar to areas where I'd worked on that project. For example, the subspecies of deer (*dakotensis*) that occupied the Black Hills region was similar in size to those (*macrourus*) I studied in Oklahoma and Arkansas (although smaller than the deer [*borealis*] I studied in Maine). The region also was punctuated with trees that were at times covered with about as much arboreal lichen as is found in the balsam fir (*Abies balsamea*) and white spruce (*Picea glauca*) forests of northern Maine (fig. 1.2).

Because of the interest of wildlife managers in the status of white-tailed deer in the Black Hills (herds had declined since the late 1970s), over my first years at SDSU, new projects were funded in the north, central, and southern regions of the "Hills." The projects were focused on habitat selection and nutrition of these deer populations, in part because of the transition of the region from aspen (*Populus tremuloides*) to ponderosa pine (*Pinus ponderosa*) (fig. 1.3), which was thought to affect population quality. I traveled many times to the region to meet with wildlife officials and USDA Forest Service employees and to help students ensure that their project goals were met. It was in early 1998 that I met with South Dakota Game, Fish and Parks employees to discuss projects that would begin in fiscal year 1999. There was some interest in conducting surveys for mountain lions because of an increase in documented sightings. And there was interest in learning about the status of the pine marten (*Martes*

FIGURE 1.2. White spruce–ponderosa pine forest of the northern Black Hills. Some trees have arboreal lichen hanging from their branches, as do balsam fir and white spruce forests of northern Maine. *Photo by Gail Jenks.*

FIGURE 1.3. Castle Creek Valley, shown in 1874 during the Custer reconnaissance of the Black Hills (*left*) and more recently (*right*). The photos show the transition from an aspen- to a pine-dominated landscape. *Progulske 1974; photos by permission from South Dakota State University.*

americana), a species formerly extinct in the region that had recently been reintroduced by means of releases of marten from western states. I proposed a study that would address both needs, which was supported and subsequently funded.

Large mammals generally have a large following of interested recreationalists, managers, and naturalists, as well as those who devote their lives to scientific study of these "charismatic megafauna." The interest may be focused on obtaining a trophy set of horns or antlers (depending on the species), or maybe on the breathtaking experience of, for example, viewing a herd of elk on a hillside. For scientists, the interest is likely to center on learning how a species interacts with its environment or what factors affect its existence. Predators, unlike deer or other large herbivorous species, are viewed both positively and negatively by various people, and those of large size, such as the mountain lion (*Puma concolor*), are both loved and despised by humans. These divergent views of mountain lions may arise just from the name "lion," which brings to mind the large cats of Africa and the Serengeti Plains, but the love-hate mix also has to do with the tabbylike appearance of kittens, the stealthy behaviors of adults that allow them to remain hidden in what might seem like open habitat, and the adult capability of killing prey of various sizes and strengths. I have had individuals call me to ask whether they can raise mountain lion kittens in their homes as they would domestic cats; others tell me they would shoot mountain lions on sight or avoid outdoor activities if any chance of encounter was even a thought. Even Theodore Roosevelt ([1885] 2004, 651), when writing about the species, stated, "It is itself a more skillful hunter than any human rival. . . . It is a beast of stealth and rapine; its great velvet paws never make a sound, and it is always on watch whether for prey or for enemies, while it rarely leaves shelter even when it thinks itself safe." Such a description likely scared most readers of his work, and it may have negatively affected the fate of the species, since at that time the mountain lion was considered vermin. Despite or maybe because of such a definition, the mountain lion was chosen as the mascot of the "Rough Riders" (a volunteer US cavalry of the Spanish-American War) (fig 1.4).

The presence of mountain lions has been known to European settlers of the new world for more than 500 years, at least since the time when Columbus first noted their existence in Honduras and Nicaragua (Young and Goldman 1946), and the species has long been an important figure in Native American culture (Young and Goldman 1946; Logan and Sweanor 2001). The adaptability of the mountain lion is represented by its historical distribution, which is one of the largest for terrestrial mammals; it has ranged from the southern tip of Chile to the Yukon Territory (Logan and Sweanor 2000). This range of presence and adaptability across the landscape may have contributed to the diversity of names (catamount, cougar, puma, panther, in addition to mountain lion) used to refer to the species.

Mountain lions were present in South Dakota historically, and in the late 1800s they were documented throughout the state (Young and Goldman 1946; Turner 1974).

FIGURE 1.4. The Rough Riders with their mountain lion mascot "Josephine." *"McClintock 1864–1934," photographic collections, US Army, Phoenix Public Library.*

Regarding the Dakota mountain lions, Roosevelt ([1885] 2004, 658) wrote: "Though the mountain lion prefers woodland it is not necessarily a beast of dense forests only, for it is found in all the plains country, living in the scanty timber belts which fringe the streams, or among the patches of brush in the Bad Lands." Roosevelt's reference to mountain lions in the plains might have come from personal observations; around that time a few specimens were captured in the oak groves along Oak Creek (where the Grand River joins the Missouri River, in Corson County, South Dakota) (Hoffman 1877). In addition, with the recolonization of the region by the species, recent observations include the plains and buttes west of Corson County and juniper stands along the Cheyenne and Missouri rivers.

In the Black Hills, members of the expedition of 1874 led by Lt. Col. George Armstrong Custer, while completing a reconnaissance of the Black Hills, reported seeing a mountain lion near the headwaters of Castle Creek (in what is now Pennington County); they believed mountain lions were numerous in the region (Ludlow 1875). Hallock (1880) also stated that mountain lions were abundant in the Black Hills at that time. However, Dodge (ca. 1875; published 1998) mentioned that only "a few" mountain lions inhabited the Black Hills. He related that when he visited the Black Hills, two or three had been seen by members of his party, but none had been killed, because the animal was rarely seen during daylight. It prowled about under darkness when they approached campsites, and because of this trait, he labeled the animal "a most arrant coward."

Not long after these initial expeditions to the Black Hills region, a bounty was placed on the mountain lion (in 1889), and by the early 1900s, the animal was considered extirpated from the state, although this was never confirmed. No official reports of mountain lion sightings in the South Dakota plains occurred at that time, and in the Black Hills, only three mountain lions were killed in the early and mid 1900s. One mountain lion was killed in 1930 at the headwaters of Stockade Beaver Creek, in Weston County, Wyoming (the west side of the Black Hills region). Another, a female, was killed in 1931, 8 km (5 miles) south of Hardy Ranger Station in Pennington County, near the South Dakota–Wyoming border. The last reported killing of a mountain lion during the bounty period in the Black Hills occurred in 1958; a male mountain lion was killed on Elk Mountain in Custer County, also along the border with Wyoming (Turner 1974). After 1958, and until the bounty was removed, verified mountain lion reports were rare. From 1964 to 1965, four mountain lions were sighted in Wind Cave National Park, and in 1965 tracks west of the town of Custer (the southern Black Hills) were positively identified as mountain lion; a mountain lion was also sighted southwest of Hot Springs, South Dakota (Fall River County). In 1968 a mountain lion was sighted near Big Crow Peak (Lawrence County, South Dakota) (Turner 1974) (fig. 1.5).

FIGURE 1.5. One of the many mountain lions captured in the Black Hills region. *Photo by Dan Thompson.*

Few documented sightings of the species occurred during the 1970s in the Black Hills region. These limited sightings contributed to the 1978 grant of legal protection in South Dakota. At that time the mountain lion was reclassified as "state threatened." The 1980s also saw few sightings, either because few animals moved into South Dakota from the west or because of the failure to report sightings during this period. However, in the early 1990s a trapper captured what was considered a relatively young male south of Selby, South Dakota (east of Mobridge and east of the Missouri River). The animal was transferred to the South Dakota Department of Game, Fish and Parks, which fitted it with a radio collar and released it in the Black Hills. The young lion (at the time estimated as 1.5 to 2.5 years old) was hit by a car not long after its release, which resulted in loss of contact with the cat; it was suspected that the radio transmitter had been damaged. However, the South Dakota Department of Game, Fish and Parks continued to receive reports of a collared mountain lion that ranged throughout the Black Hills region, and in 1998 the male was killed in Custer County. The cat was transferred to South Dakota State University, where necropsy confirmed that the vehicle incident likely caused transmitter malfunction but also blindness in one eye and damage to a front leg. At the time of its death, the cat appeared to be malnourished and had been consuming porcupine (*Erethizon dorsatum*), likely because of its injured condition as well as the ease of capture of this common prey species.

It is unclear whether the mountain lions sighted in the 1960s had immigrated to the Black Hills or whether they were descendants of a few individuals remaining on the mountain range (Turner 1974), or both (although there were no observations of kittens at that time). The nearest mountain ranges to the Black Hills are the Bighorn Mountains, 200 km (125 miles) to the west, and the Laramie Mountains, about 160 km (100 miles) to the southwest (fig. 1.6). Berg, McDonald, and Strickland (1983) believed that transient mountain lions originating from established populations in the Bighorn Mountains recolonized the Black Hills. During our studies, one radio-collared lion did travel from the Black Hills to the Bighorn Mountains, providing support for this hypothesis. In addition, Anderson (2003) determined that the genetic structure of mountain lions from the central Rocky Mountains in Wyoming was similar to that of the Black Hills population, indicating that gene flow occurred among mountain lions occupying five mountain ranges in the two states and further supporting the movement of lions among these mountain ranges. Because the topographic orientation of major draws originating in southeastern Wyoming are southwest to northeast, ending in the southern Black Hills region, and those originating between the Bighorn Mountains and the Black Hills have a similar orientation, ending in the North Dakota Badlands region, landscape characteristics would facilitate southwest-to-northeast dispersal and colonization of the southern Black Hills.

In support of this belief about the movement of mountain lions into the Black Hills, in 1995 Ted Benzon, senior big game biologist, South Dakota Department of Game, Fish and Parks, began recording sightings of mountain lions in the Black Hills. He

FIGURE 1.6. Locations of other mountain ranges with established mountain lions populations west of the Black Hills region. *Figure by Dan Thompson.*

observed an overall increase in reported sightings from 1995 to 1999, but those sightings were not randomly distributed in the Black Hills. When the numbers of reported sightings were adjusted for county population size, more reports were obtained from the southern counties (Custer and Fall River counties) than the northern counties (Lawrence and Pennington counties) (fig. 1.7). In addition, from 1996 to 1999 more mountain lion deaths were reported in the southern two counties (58%) than in the counties of the northern Black Hills (42%) (Fecske, Jenks, and Lindzey 2003). This was the period when I was involved in studies of white-tailed deer in the Black Hills. During those studies, which began in the northern (early 1990s) and extended to the central (mid 1990s) Black Hills, no mortalities of radio-collared deer were categorized as resulting from mountain lion predation, although a mountain lion sighting occurred near Deerfield Lake in the central Black Hills during the mid 1990s.

When we began our work on the mountain lion, Chuck Anderson was conducting a study (Anderson 2003) that genetically compared mountain lions throughout Wyoming. He had helped train my first graduate student, Dorothy (Fecske) Wells, who studied mountain lions and asked for blood samples to add to his study. We supplied a few samples from our first captures in the Black Hills, which had a slightly lower

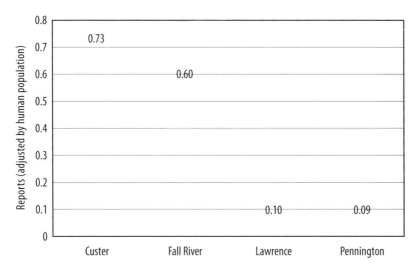

FIGURE 1.7. Mountain lion reports (sightings) in the Black Hills, adjusted for county population size. *Skelton and Jenks 2000.*

heterozygosity (H_{obs}) (0.49) than four other geographically distinct mountain lion populations (H_{obs} 0.51–0.56) located in Wyoming (Anderson, Lindzey, and McDonald 2004); the mountain lion samples from the Black Hills used in this analysis were established males and females captured in the southern Black Hills (see chapter 7 for recent estimates of heterozygosity for Black Hills lions). Anderson's findings seemed to support the low number of individual mountain lions in the Black Hills region at the time. Low population size is generally associated with reduced genetic variability, and in the Black Hills the estimated population size at that time was "just a few" mountain lions. Therefore, most evidence pointed to the hypothesis that there were indeed only a few animals in the Black Hills and that, because most were likely located in the southern region, mountain lions likely had recolonized the Black Hills from the southwest.

Nevertheless, after mountain lions received protection, sightings began increasing, especially in the southern region of the Black Hills. In 1997, based on anecdotal information, the South Dakota Department of Game, Fish and Parks estimated that there were 15 to 25 mountain lions in the Black Hills (SDGFP 1998b). The agency became concerned about mountain lions after a few incidents occurred involving depredation to livestock and horses and after two kills (mule deer, *Odocoileus hemionus*) (fig. 1.8) made by mountain lions were found within the Rapid City limits (T. Benzon, South Dakota Game, Fish and Parks, pers. commun.); Rapid City, on the outskirts of the Black Hills, had at the time a population of 57,513 (as estimated by the Rapid City Chamber of Commerce in 2003). Because livestock is a major commodity in the region, and mountain lions had attacked people in other western states (Beier 1991), in 1998 the South Dakota Department of Game, Fish and Parks drafted a mountain lion action

FIGURE 1.8. A deer is killed (*top*) and cached (*bottom*) by a mountain lion in the Black Hills. *Photos by Rebeca Juarez.*

plan (SDGFP 1998a), and cooperative research efforts were initiated with SDSU to learn more about the species in the Black Hills. In 2003, based on our initial study of mountain lions in the Black Hills (Fecske 2003) and ongoing work on the species (Thompson 2009), the species was removed from the list of state-threatened species and reclassified as a big game animal with a closed season, to be managed by the South Dakota Department of Game, Fish and Parks (Fecske 2003; Anderson, Lindzey, and McDonald 2004). This relisting was based in part on initial research findings indicating that there were more mountain lions in the Black Hills than originally thought. In addition, some of the mortalities of radio-collared mountain lions were due to interactions with hunters and trappers, who had stated they were threatened by the species and had to defend themselves.

At the time, it was believed that reclassifying the species as a big game animal provided more protection than the state-threatened status. In fact, one of the first mountain lions radio-collared in the southern Black Hills (around Battle Mountain, just south of Wind Cave National Park) was killed by a predator hunter using a traditional caller, when the lion approached the hunter in response to the call. The hunter cut the collar off the lion and threw it into the Cheyenne River. The hunter may have thought that immersing the collar in water would short-out the electronics and stop the signal emitted from the collar. However, the collars are designed for exposure to water, since wild animals, whether deer or their predators, typically cross water systems in their day-to-day activities. While flying to locate the animal, Dorothy (Fecske) Wells received the signal from the collar and retrieved it from the river. Although the species was classified "state threatened," the hunter was not fined, because he stated that he was threatened by the approach of the animal.

Historically, the management of many carnivore species was based more on art than on science (CMGWG 2005). In South Dakota, initial efforts to manage mountain lions consisted primarily of population-level assessment and removal of problem animals to address public safety or to reduce depredation. As more mountain lion–human interactions and conflicts occurred, intensive research was carried out; additional animals were tracked, more research questions were raised, and more objectives were proposed. Nevertheless, no attacks on humans were documented during this time of initial population establishment and increase, despite increases in sightings, kills of domestic species (llamas and sheep), and an increase in the recreational use of the Black Hills region.

In 2005, based on information collected on mountain lions during the first few years of study, the South Dakota Department of Game, Fish and Parks initiated a mountain lion harvest season. The first season was limited to a harvest of 12 animals, and no more than half could be female. The announcement of this first harvest season resulted in a lawsuit, initiated by the Mountain Lion Foundation of California and supported by the Black Hills Mountain Lion Foundation, seeking an injunction to stop the harvest. The justification for the suit was, first, the limited amount of knowledge available on the species at that time and, second, that because of the isolated nature of the Black Hills and the suspected low population size of newly established mountain lions, such a season likely would cause extinction of the species in the Black Hills. The case was heard in Pierre, South Dakota, a week before the season was set to begin on October 1. Based on the evidence provided by the State of South Dakota, which included additional information about the population, gathered since Dorothy (Fecske) Wells completed the first project on the species in 2003—more than 30 mountain lions had been radio-collared by the time of the hearing—the judge ruled in support of the state, and the harvest began. A harvest of mountain lions has continued in the Black Hills since 2003, and harvest limits have been increased until recently.

Since beginning our studies of mountain lions back in the late 1990s, my students and I have followed the recolonization of the Black Hills by the species from a period when just a few lions were believed to inhabit the region, through a period when the species was believe to be "saturated," resulting in adult and kitten starvation (fig. 1.9) and increased road mortality, through periods of nonhunting and hunting as well as periods of enthusiasm for the species (when the general public was filled with excitement that the ecosystem was again "complete" and sightings of the species were

FIGURE 1.9. Emaciated mountain lion found dead in the Black Hills during a period when the population was believed to be saturated. *Photo by Steve Griffin.*

greeted with fascination), to a period of hatred focused on mountain lions because the number of large prey (deer, elk [*Cervus elaphus*], and bighorn sheep [*Ovis canadensis*]) killed by this large predatory species was believed to result in reduced hunting opportunities for these as well as other species occupying the region. Over such a span of time, covering significant changes in how mountain lions used the Black Hills and were perceived by the public, we were able to gain insight into the ecology and general view of this charismatic species.

The Black Hills provide a unique system for the study of mountain lions because the region is semi-isolated from other areas inhabited by mountain lions, and the prairies that surround the ecoregion represent an alternative habitat with which established lions and their young are unfamiliar. At times during our studies, we were able to view lions leaving the ecoregion. More than once did lions, which we expected to disperse from the region because of their age, venture away (mostly to the west) only to move back to the mountain range at a later time. We suspect that these lions were in search of habitats that resembled their natal area (ponderosa pine forest), and in fact many dispersing lions traversed the boundaries of the ecoregion before leaving to the northwest. This northwest area may lend itself to dispersal and movements out of the region because of the ponderosa pine fingers of habitat that extend to the Bear Lodge Mountains and pockets of Custer National Forest.

The following chapters provide an overview of information collected while studying this mountain lion population over a 17-year period. During this time we focused our activities on collecting information to answer questions on population size, movements, genetics, and prey use of the species. Although some of our work was limited to specific time intervals, the long-term effort provided opportunities to describe relationships between population size and other factors, such as survival, home range size, and dispersal, as well as the effect of harvest and how the public perceived the species.

Literature Cited

Anderson, C. R. 2003. Cougar ecology, management, and population genetics in Wyoming. PhD diss., University of Wyoming.

Anderson, C. R., Jr., F. G. Lindzey, and D. B. McDonald. 2004. Genetic structure of cougar populations across the Wyoming Basin: Metapopulation or megapopulation. Journal of Mammalogy 85:1207–1214.

Beier, P. 1991. Cougar attacks on humans in the United States and Canada. Wildlife Society Bulletin 19:403–412.

Berg, R. L., L. L. McDonald, and M. D. Strickland. 1983. Distribution of cougars in Wyoming as determined by mail questionnaire. Wildlife Society Bulletin 11:265–270.

Cougar Management Guidelines Working Group (CMGWG). 2005. Cougar Management Guidelines. 1st ed. Wild Futures, Bainbridge Island, WA.

Dodge, R. I. 1998. The Black Hills. Gretna, LA: Pelican.

Fecske, D. M. 2003. Distribution and abundance of American martens and cougars in the Black Hills of South Dakota and Wyoming. PhD. diss., SDSU.

Fecske, D. M., J. A. Jenks, and F. G. Lindzey. 2003. Characterisitics of mountain lion mortalities in the Black Hills, South Dakota. In Proceedings of the sixth Mountain Lion Workshop, 25–29: Texas Parks and Wildlife Department.

Hallock, C. 1880. The sportsman's gazetteer and general guide. 5th ed. New York: Forest and Stream.

Hoffman, W. J. 1877. List of mammals found in the vicinity of Grand River. Proceedings of the Boston Society of Natural History 19:94–102.

Jenks, J. A., and D. M. Leslie Jr. 1988. Effect of lichen and in vitro methodology on digestibility of winter deer diets in Maine. Canadian Field-Naturalist 102:216–220.

Jenks, J. A., and D. M. Leslie Jr. 1989. Digesta retention of winter diets in white-tailed deer (Odocoileus virginianus) fawns in Maine, USA. Canadian Journal of Zoology 67:1500–1504.

Logan, K. A., and L. L. Sweanor. 2000. Puma. In Ecology and Management of Large Mammals in North America, ed. S. Demarais and P. R. Krausman, 347–377. Upper Saddle River, NJ: Prentice Hall.

Logan, K. A., and L. L. Sweanor. 2001. Desert Puma: Evolutionary ecology and conservation of an enduring carnivore. Washington, DC: Island Press.

Ludlow, W. 1875. Report of a reconnaissance of the Black Hills of Dakota made in the summer of 1874. Washington, DC: Government Printing Office.

Progulske, D. R. 1974. Yellow ore, yellow hair, yellow pine: A photographic study of a century of forest ecology. Brookings: Agricultural Experiment Station, SDSU.

Roosevelt, T. (1885) 2004. Hunting trips of a ranchman, and The wilderness hunter. Reprint, New York: Modern Library.

Skelton, W. R., and J. A. Jenks. 2000. Temporal changes in mountain lion, Puma concolor, Reports in the Black Hills of South Dakota. Undergraduate Research Poster Session, State Capitol, Pierre, SD.

South Dakota Department of Game, Fish and Parks (SDGFP). 1998a. Action plan for managing Cougar/human/property interactions in western South Dakota. South Dakota Division of Wildlife, Pierre, SD.

South Dakota Department of Game, Fish and Parks (SDGFP). 1998b. Yes, a few cougars live in South Dakota. South Dakota Department of Game, Fish and Parks, Pierre, SD.

Thompson, D. J. 2009. Population demographics of cougars in the Black Hills: Survival, dispersal, morphometry, genetic structure, and associated interactions with density dependence. PhD diss., SDSU.

Turner, R. W. 1974. Mammals of the Black Hills of South Dakota and Wyoming. Miscellaneous Publication No. 60, University of Kansas Publications. Lawrence: Museum of Natural History, University of Kansas.

Young, S. P., and E. A. Goldman. 1946. The puma: Mysterious American cat. New York: Dover.

CHAPTER 2

Ecology of the Black Hills

The Black Hills are located about 640 km (about 400 miles) due west of Brookings and SDSU (fig. 2.1). Because many of my initial research projects were focused in the Black Hills, most of my trips to visit students combined business and pleasure. My wife, Gail, survived shuttles of students to the Black Hills, meetings with state officials or with foundations where I was serving as a committee member, or just the rescheduling of a recreational adventure that was delayed owing to project needs. There were times when meetings developed just because I happened to be "in the area." In one instance, we transported a student from Brookings to Rapid City, where she stayed with another student while finding a place to live and assembling survey equipment stored at the Rapid City office of the South Dakota Department of Game, Fish and Parks. After completing the transfer, we traveled to the middle of the Black Hills and spent a few enjoyable days camping, hiking, and fishing at Deerfield Lake. On another occasion, we brought two of our grandchildren with us to introduce them to the region. While I attended a Project Advisory Committee Meeting for the Rocky Mountain Elk Foundation, they experienced some of the commercial venues (such as Reptile Gardens) that are found just outside of Rapid City.

The trip from Brookings to the Black Hills generally takes about six hours, as the Black Hills region straddles west-central South Dakota and northeastern Wyoming. Geologists consider the range an eastern extension of the Rocky Mountains and, as such, one of the oldest mountainous regions in North America, if not the oldest (Froiland 1990). The region is elliptically shaped and oriented in a north-south direction. One reason the Black Hills were left unexplored until Custer's 1874 reconnaissance of the region was that, in the Treaty of 1851, the lands surrounding the range had been ceded to the Sioux Nation, who referred to the region as "Paha Sapa." After the reconnaissance, with the signing of the Treaty of 1876, the Sioux relinquished claim

FIGURE 2.1. View from the Hogback surrounding the Black Hills. *Photo by Gail Jenks.*

to the Black Hills region (Progulske 1974). This later treaty also created a safe travel zone for settlers and prospectors interested in panning for gold from eastern South Dakota through the central region of western South Dakota (fig. 2.2).

The geology of the Black Hills also marks the mountain chain as one of the oldest regions on the North American continent. Portions of the Black Hills date as early as the Precambrian Period to as late as the Pliocene Period. They were formed by a process known as "updoming." As a dome of earth was pushed upward, the top eroded away, leaving exposed rock layers. Soils of the Black Hills are identified as in the gray wooded soil region, which is unique for South Dakota. These soils were largely developed under timber in dry subhumid to humid climate and are derived from limestone, sandstone, and local alluvium via igneous and metamorphic rocks (Froiland 1990). The rockiness of the region provides unlimited crevasses (e.g., Spring Creek and Spearfish Canyon) where mountain lions can escape notice, evade pursuit by hunters as well as researchers, and use stealth to capture the diverse prey species that inhabit the region.

Major topographical features of the range include a hogback range that encircles the Black Hills (fig. 2.1); the Red Valley, which separates the hogback from the Black Hills proper; the Limestone Plateau, which rises more than 2,000 m (6,560 ft.) within the central region of the mountains; and the Harney Range, which also is centrally situated within the Black Hills (Froiland 1990). The Black Hills extend about 190 km (120 miles) from north to south and 95 km (60 miles) from east to west (Petersen 1984);

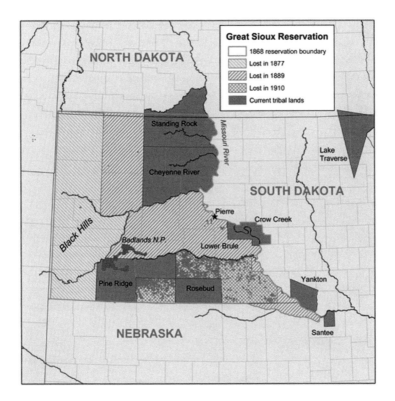

FIGURE 2.2. Map of the Sioux reservation, created March 18, 2008.
Karl Musser. See https://commons.wikimedia.org/wiki/File:Siouxreservationmap
.png#/media/File:Siouxreservationmap.png.

the highest point, at Harney Peak (now Black Elk Peak), reaches 2,207 m (7,240 ft.) and towers 1,500 m (4,920 ft.) above the surrounding prairies. The mountain range is isolated, being completely surrounded by the Northern Great Plains and covering approximately 8,400 km² (3,240 miles²) (Fecske, Jenks, and Smith 2002). Like the surrounding prairies of South Dakota that are west of the Missouri River, the region was unglaciated; the Missouri River, centrally located in South Dakota, is about 240 km (150 miles) east of the Black Hills (Froiland 1990; see fig. 2.2).

The Black Hills ecosystem is diverse and comprises four distinct vegetation complexes: Rocky Mountain coniferous forest, northern coniferous forest, a grassland complex, and a deciduous complex. When Custer visited the region in 1874, the primary forest cover was quaking aspen (*Populus tremuloides*). Presently, the forest cover in the Black Hills is predominantly ponderosa pine (*Pinus ponderosa*), with white spruce (*Picea glauca*) and aspen occurring predominately above the 2,000 m contour (on the limestone plateau) (fig. 2.3).

FIGURE 2.3. Vegetation of the Black Hills. In the image, black vegetation is primarily ponderosa pine and white spruce. Lighter-colored vegetation is a mix of grasslands, burns, shrublands, deciduous trees, and open water. *Fecske 2003.*

The deciduous complex characterizing the Black Hills is dominated by a mixture of streamside trees, including American elm (*Ulmus americana*), green ash (*Fraxinus pennsylvanica*), box elder (*Acer negundo*), and eastern hop-horn-beam (*Ostrya virginiana*) in the lower elevations; those trees are gradually replaced by aspen and paper birch (*Betula papyrifera*) at higher elevations. Bur oak (*Quercus macrocarpa*) forests occur in warm, dry, low-elevation sites represented by finger draws that are found along the edges of the Black Hills (Hoffman and Alexander 1987). Understory plants of the region are diverse (fig. 2.4) and include ground juniper (*Juniperus horizontalis*), Oregon grape (*Berberis repens*), and bearberry (a.k.a. Kinnikinnick, *Arctostaphylos uva-ursi*) in the north; snowberry (*Symphoricarpus albus*) and serviceberry (*Amelanchier alnifolia*) are common in the southern Black Hills. Common grasses include carices (*Carex* spp.), poverty oat grass (*Danthonia spicata*), and rough rice grass (*Oryzopsis*

FIGURE 2.4. Ear-tagged white-tailed deer utilizing the thick understory vegetation that characterizes the northern Black Hills. *Photo from Department of Wildlife and Fisheries Sciences, South Dakota State University.*

asperifolia). Common forbs include bluebell (*Campanula rotundifolia*) (a favorite forage of elk), goldenrod (*Solidago* spp.), western yarrow (*Achillea millefolium*), and vetchling (*Lathyrus ochroleucus*) (Gibbs et al. 2004). Overall, about 1,000 plant species occur in the region (Larson and Johnson 1999).

Mammals historically documented as being killed and consumed by mountain lions in the Black Hills region include large prey species such as white-tailed deer (fig. 2.4), mule deer, elk, bighorn sheep, and mountain goat (*Oreamnos americanus*). Other smaller (i.e., meso- and micromammal) species include porcupine (*Erethizon dorsatum*), coyote (*Canis latrans*), marten (*Martes americana*), and yellow-bellied marmot (*Marmota flaviventris*), as well as small mammals such as voles (i.e., southern redbacked, *Myodes gapperi*) (Smith 2002; Smith et al. 2002). At times, mountain lions kill and consume domestic livestock species such as llamas, goats, sheep, donkeys, turkeys, and domestic dogs and cats in the region.

Despite the diverse prey assemblage available to mountain lions in the Black Hills system, prey are not uniformly distributed. Deer were first documented in the region in 1874 by George Grinnell, who accompanied General Custer during the reconnaissance of the Black Hills. "White-tails" are found through the region, whereas mule deer are generally located around the fringes of the Black Hills; populations total more than 40,000 animals (Cudmore 2017). White-tails outnumber mule deer by about

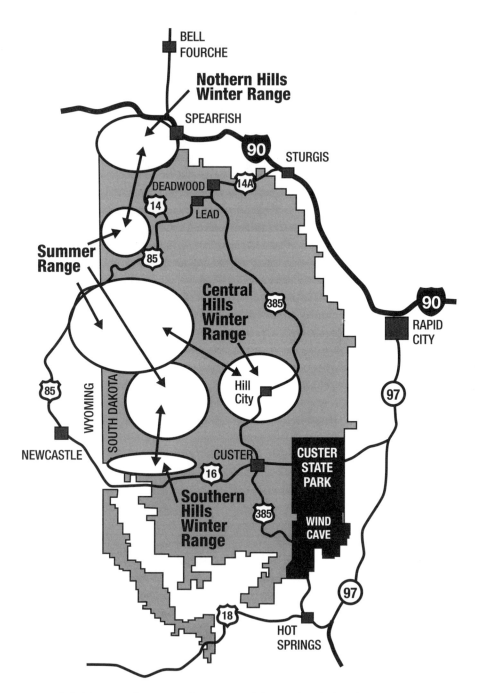

FIGURE 2.5. Migrational patterns of white-tailed deer in the Black Hills. *Adapted from DePerno et al. 2002.*

4:1–5:1 in this area (Turner 1974). Furthermore, white-tailed deer are migratory, with regions displaying unique migratory patterns; northern white-tails migrate about 16 km north-south (their high-elevational summer range is south of their low-elevational winter range [the northern fringe of the Black Hills]), central white-tails migrate 48 km northwest-southeast (their summer range is located within and along the Wyoming border), and southern white-tails migrate north-south (their summer range is north of their low-elevational winter range) (Griffin et al. 1994, 1999; Griffin, Jenks, and DePerno 2004; fig. 2.5). Because white-tails are the most common prey available to mountain lions in the Black Hills, we originally hypothesized that the diverse migrational patterns of white-tails could influence the movements and patterns of occupancy of this large predator.

Elk and bighorn sheep (fig. 2.6), historical inhabitants of the region, have been reintroduced, along with mountain goats. Elk mainly occupy the northern, west central, and southern regions of the Black Hills, while bighorn sheep herds occur in the Rapid City, Spring Creek, Custer State Park, and Elk Mountain areas (Parr 2015). Mountain goats occur around Black Elk Peak and Mount Rushmore. Pronghorn (*Antilocapra americana*) are limited to the Custer State Park and Wind Cave National Park area but have been sighted in other areas of the Black Hills.

The predator guild of the Black Hills also includes both coyote and bobcat (*Lynx rufus*). Historically, the region was occupied by gray wolf (*Canis lupus*) and grizzly

FIGURE 2.6. Bighorn sheep have been reintroduced to the Black Hills. *Photo by Brynn Parr.*

FIGURE 2.7. General George Armstrong Custer killed a grizzly bear during the reconnaissance of the Black Hills in 1874. *Progulske 1974.*

bear (*Ursus arctos*) (fig. 2.7). Interactions between predators have been documented within the Black Hills region. For example, mountain lions have killed and partially consumed coyotes (Smith 2014), both in the Black Hills and in other regions, for example, the North Dakota Badlands (Wilckens et al. 2016). Because of the relatively high density of bobcats in the region, killing likely occurs between lions and bobcats as well, but there is no evidence that the harvest of bobcats was affected by the recent increase in mountain lions in the region.

The Black Hills National Forest (BHNF) is highly developed and is one of the most heavily roaded national forests, with 13,411 km of inventoried roads within and adjacent to the BHNF boundary (USDA Digital Line Graph Files, Black Hills National Forest Service Database, Custer, SD, unpublished data; fig. 2.8). About 90% of the Black Hills is surrounded by four-lane highways, and overall road density for the Black Hills National Forest–Custer State Park region is estimated at 2.1 km road/km^2 (fig. 2.8; D. M. Fecske, unpublished data). Also, the Black Hills National Forest has a higher

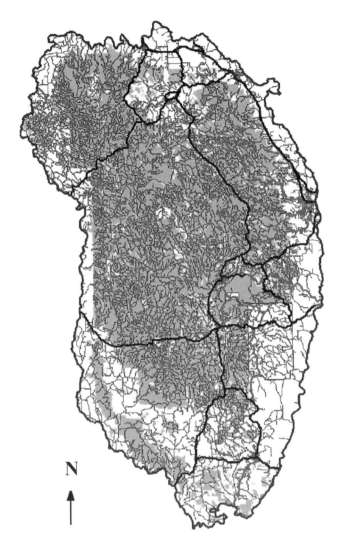

N

↑

FIGURE 2.8. Extent of roads in the Black Hills National Forest region of the Black Hills. Dark (black) roads represent highways (e.g., Interstate 90). Other roads (gray) illustrate the high density of roads within the Black Hills National Forest. *Fecske 2003; Baker and Knight 2000.*

road density than do nearby national forests in Wyoming and Colorado (Baker and Knight 2000).

Fecske (2003; fig. 2.9), constructed a digital habitat-relation model to help determine the distribution of mountain lions in the Black Hills National Forest. Mountain lion habitat was characterized by vegetation and topography in which prey were

FIGURE 2.9. Classification of mountain lion habitat found in the Black Hills National Forest. Habitat was classed based on prey availability (white-tailed deer), slope, and terrain. Darker habitats have lower suitability for mountain lions. *Fecske 2003.*

abundant; those features provide stalking and concealment cover for the species, increasing the chances that mountain lions can make successful kills (Seidensticker et al. 1973). Certain species of vegetation were not included in this mapping because they were considered important to mountain lions only in that they attract prey, predominantly deer and deer-sized ungulates; but vegetation that provides food and cover for prey also provides stalking cover for mountain lions (Logan and Sweanor 2000). For example, in the Black Hills, we observed a mountain lion stalking white-tailed deer using dense chokecherry (*Prunus virginiana*) shrubs for concealment; the chokecherry bushes provided the same herd of deer a food source, especially in winter. We have also located both radio-collared mountain lions and one uncollared mountain lion in dense white spruce (*Picea glauca*) stands (Fecske 2003). We hypothesized that the animals were using the cool, moist stands for concealment cover, thermoregulation during hot weather (Jalkotzy, Ross, and Wierzchowski 1999), and stalking cover, since white-tailed deer in the Black Hills are known to select mixed white spruce stands during summer (DePerno 1998).

Dense vegetation also is used by mountain lions for concealment and security cover while feeding and as nursery sites by female mountain lions (Logan and Sweanor 2000); in the Black Hills dense vegetation often occurs along drainages (Klaver et al. 2008) (fig. 2.10). Steep slopes, boulder piles, undercut cliffs, and rock outcrops serve mountain lions much as vegetation does. In turn, rugged terrain is important to ungulates seasonally, as exposed south- and west-facing slopes provide food and mild weather conditions in winter (Logan and Sweanor 2000). In the northern Black Hills, north-facing slopes also provide these same benefits for deer (Kennedy 1992). Logan and Irwin (1985) suggested that density estimates of mountain lions obtained from areas with specific habitat features could be used to estimate breeding population densities in areas with similar habitat. Home ranges also have been used to determine density estimates of large carnivores (Clark, Dunn, and Smith 1993). We used our habitat-relation model to identify high-quality habitat to predict mountain lion occurrence. In addition, in conjunction with data collected from a sample of radio-collared mountain lions, the literature, and a mountain lion population program (Program PUMA; Beier 1993), we were able to generate an initial population estimate for the region (Fecske 2003). Our projected population indicated that more than 100 mountain lions could be occupying the Black Hills, but when I presented these results at a regional meeting, my suggestion met with considerable objections from the audience; in fact, many in the audience responded with disbelief.

Seasonal temperature fluctuations in the Black Hills are characteristic of the continental climate that is typical of the region (Orr 1959). Mean annual temperatures range from 5°C to 9°C with extremes of −40°C to −44°C. Mean annual precipitation is greater than 66 cm (Orr 1959), and annual snowfall can exceed 254 cm at higher elevations (Thilenius 1972); snow in the southern regions of the Black Hills is unpredictable and rarely lasts more than a few days. The region is known to receive late winter storms with significant snowfall. In fact, migrating whitetails have died during these storms, likely owing to lack of forage and increased susceptibility to predators (DePerno et al. 2000). Relative to climate, the Black Hills can be separated into northern and southern regions (Froiland 1990). The northern area is characterized by colder temperatures, deeper snows, and higher precipitation, whereas the southern

FIGURE 2.10. Typical thick vegetation that occurs in the Black Hills. *Photo by D. M. Fecske.*

FIGURE 2.11. Example of habitat burned in the southern Black Hills. *Photo by Gail Jenks.*

region is warmer, receives about 25 cm less in precipitation, and generally has more sunny days.

Because of the variation in precipitation, both across the region and annually, fires are commonplace. For example, in 2000 the Jasper fire, a high-severity fire, burned approximately 33,729 ha (about 7% of the Black Hills) (Jasper Fire Rapid Assessment, www.fs.fed.us/r2/blackhills/fire/history/jasper), from August until official containment was declared in September 2000. Fires like the Jasper fire can be highly variable, resulting in a mosaic of areas burned interspersed with areas with low-intensity burns and unburned areas (fig. 2.11; Dubreuil 2002). In addition, grassy meadows interspersed with aspen (*Populus tremuloides* Michx.) and paper birch (*Betula papyrifera* Marsh.) naturally occur throughout the pine forest in bottom draws. An example, Gillette Prairie, is located in the central region of the mountains and provides forage for a diverse assemblage of ungulates, including elk and pronghorn, that occupy the area (Simpson 2015).

Three areas within the Black Hills where mountain lions are federally protected are under the administration of the National Park Service: Mount Rushmore National Memorial (about 6 km^2), Jewel Cave National Monument (about 5 km^2), and Wind Cave National Park (about 110 km^2). Custer State Park (about 286 km^2) is administered by the State of South Dakota through the South Dakota Department of

Game, Fish and Parks. During the first years of harvest, mountain lions were protected here as well as on federal properties. The Wyoming portion of the Black Hills is currently managed to reduce the local population, and high harvest rates are prescribed (Thompson and Large Carnivore Section 2013).

Literature Cited

Baker, W. L., and R. L. Knight. 2000. Roads and forest fragmentation in the southern Rocky Mountains. In Forest fragmentation in the southern Rocky Mountains, ed. R. L. Knight, F. W. Smith, S. W. Buskirk, W. H. Romme, and W. L. Baker, 97–122. Boulder: University Press of Colorado.

Beier, P. 1993. Puma: A population simulator for cougar conservation. Wildlife Society Bulletin 21:356–357.

Clark, J. D., J. E. Dunn, and K. G. Smith. 1993. A multivariate model of female black bear use for a geographic information system. Journal of Wildlife Management 57:519–526.

Cudmore, K. W. 2017. An evaluation of deer and pronghorn surveys in South Dakota. Master's thesis, SDSU.

DePerno, C. S. 1998. Habitat selection of a declining white-tailed deer herd in the central Black Hills, South Dakota and Wyoming. PhD diss., SDSU.

DePerno, C. S., J. A. Jenks, S. L. Griffin, and L. A. Rice. 2000. Female survival rates in a declining white-tailed deer population. Wildlife Society Bulletin 28:1030–1037.

DePerno, C. S., J. A. Jenks, S. L. Griffin, L. A. Rice, and K. F. Higgins. 2002. White-tailed deer habitats in the central Black Hills. Journal of Range Management 55:242–252.

Dubreuil, R. P. 2002. Habitat selection of white-tailed and mule deer in the southern Black Hills, South Dakota. Master's thesis, SDSU.

Fecske, D. M. 2003. Distribution and abundance of American martens and cougars in the Black Hills, South Dakota and Wyoming. PhD. diss., SDSU.

Fecske, D. M., J. A. Jenks, and V. J. Smith. 2002. Field evaluation of a habitat-relation model for the American marten. Wildlife Society Bulletin 30:775–782.

Froiland, S. G. 1990. Natural history of the Black Hills and Badlands. Center for Western Studies, Augustana College, Sioux Falls, SD.

Gibbs, M. C., J. A. Jenks, C. S. DePerno, B. F. Sowell, and K. J. Jenkins. 2004. Cervid forage utilization in noncommercially thinned ponderosa pine forests. Journal of Range Management 57:435–441.

Griffin, S. L., J. A. Jenks, and C. S. DePerno. 2004. Seasonal movements and home ranges of white-tailed deer and mule deer in the southern Black Hills, South Dakota, 1998–2003. Completion Report No. 7583, Federal Aide to Wildlife Restoration, Job W-75-R-34. South Dakota Department of Game, Fish and Parks, Pierre, SD.

Griffin, S. L., J. F. Kennedy, L. A. Rice, and J. A. Jenks. 1994. Movements and habitat use of white-tailed deer in the northern Black Hills, South Dakota, 1990–1992. Completion Report, Federal Aide to Wildlife Restoration, Job W-75-R-33. South Dakota Department of Game, Fish and Parks, Pierre, SD.

Griffin, S. L., L. A. Rice, C. S. DePerno, and J. A. Jenks. 1999. Seasonal movements and home ranges of white-tailed deer in the central Black Hills, South Dakota and Wyoming,

1993–1997. Completion Report No. 99-03, Federal Aide to Wildlife Restoration, Job W-75-R-34. South Dakota Department of Game, Fish and Parks, Pierre, SD.

Hoffman, G. R., and R. R. Alexander. 1987. Forest vegetation of the Black Hills National Forest of South Dakota and Wyoming: A habitat type classification. USDA Research Paper RM-276. Rocky Mountain Forest and Range Experiment Station, Fort Collins, CO.

Jalkotzy, M. G., I. P. Ross, and J. Wierzchowski. 1999. Cougar habitat use in southwestern Alberta. Prepared for Alberta Conservation Association. Arc Wildlife Services Ltd., Calgary, Alberta.

Kennedy, J. F. 1992. Habitat selection by female white-tailed deer in the northern Black Hills of South Dakota and Wyoming. Master's thesis, SDSU.

Klaver, R. W., J. A. Jenks, C. S. DePerno, and S. L. Griffin. 2008. Associating seasonal range characteristics with survival of white-tailed deer. Journal of Wildlife Management 72:343–353.

Larson, G. E., and J. R. Johnson. 1999. Plants of the Black Hills and Bear Lodge Mountains. Brookings, SD: SDSU.

Logan, K. A., and L. L. Irwin. 1985. Cougar habitats in the Big Horn Mountains, Wyoming. Wildlife Society Bulletin 13:257–262.

Logan, K. A., and L. L. Sweanor. 2000. Puma. In Ecology and Management of Large Mammals in North America, ed. S. Demarais and P. R. Krausman, 347–377. Upper Saddle River, NJ: Prentice Hall.

Orr, H. K. 1959. Precipitation and streamflow in the Black Hills. Fort Collins, CO: Rocky Mountain Forest and Range Experiment Station.

Parr, B. L. 2015. Population parameters of a bighorn sheep herd inhabiting the Elk Mountain Region of South Dakota and Wyoming. Master's thesis, SDSU.

Petersen, L. E. 1984. Northern Plains. In White-tailed deer: Ecology and management, ed. L. K. Halls, 441–448. Harrisburg, PA: Stackpole Books.

Progulske, D. R. 1974. Yellow ore, yellow hair, yellow pine: A photographic study of a century of forest ecology. Brookings: Agricultural Experiment Station, SDSU.

Seidensticker, J. C., IV, M. G. Hornocker, W. V. Wiles, and J. P. Messick. 1973. Mountain lion social organization in the Idaho Primitive Area. Wildlife Monographs 35:1–60.

Simpson, B. D. 2015. Population ecology of Rocky Mountain elk in the Black Hills, South Dakota and Wyoming. Master's thesis, SDSU.

Smith, J. B. 2014. Determining impacts of mountain lions on bighorn sheep and other prey sources in the Black Hills. PhD diss., SDSU.

Smith, V. J. 2002. Mammal distributions and habitat models for South Dakota. Master's thesis, SDSU.

Smith, V. J., J. A. Jenks, C. R. Berry Jr., C. J. Kopplin, and D. M. Fecske. 2002. The South Dakota Gap Analysis Project. Final Report. Research Work Order No. 65. Department of Wildlife and Fisheries Sciences, SDSU, Brookings, SD.

Thilenius, J. F. 1972. Classification of deer habitat in the ponderosa pine forest of the Black Hills, South Dakota. Fort Collins, CO: US Department of Agriculture, Forest Service, Rocky Mountain Forest and Range Experiment Station.

Thompson, D. J., and Large Carnivore Section. 2013. Wyoming mountain lion harvest/mortality report: Harvest years 2010–2012 (1 September 2010–15 April 2012). Wyoming Game and Fish Department, Lander, WY.

Turner, R. W. 1974. Mammals of the Black Hills of South Dakota and Wyoming. Miscellaneous Publication No. 60, University of Kansas Publications. Lawrence: Museum of Natural History, University of Kansas.

Wilckens, D. T., J. B. Smith, S. A. Tucker, D. J. Thompson, and J. A. Jenks. 2016. Mountain lion (*Puma concolor*) feeding behavior in the recently recolonized Little Missouri Badlands, North Dakota. Journal of Mammalogy 97:373–385.

CHAPTER 3

Characteristics of Black Hills Mountain Lions

When we set out to capture mountain lions in 1999, few thought we would be successful, because of the low number of individuals believed to inhabit the Black Hills and the rather complex methods needed to accomplish the task. One of our first successful captures occurred along the South Dakota–Wyoming border. We had come upon fresh tracks just after a light snow and assembled a crew, including a houndsman, employees of the South Dakota Department of Game, Fish and Parks, and my PhD student Dorothy (Fecske) Wells; because this was about our fourth attempted capture, I traveled to the region to assist.

We headed out to the area where the tracks had been found (by the houndsman) and were able to tree the approximately 3.5-year-old male (M4) in short order, not far from a road on private land (fig. 3.1). Luckily, the landowner was interested in the project, and once we had the lion sedated, he and his wife came out to the capture site, which was not far from their home. They watched while we collected biological data, including measurements, body weight, age, and general condition (i.e., presence of parasites, injuries, scars), and radio-collared the animal. The male lion was in superb condition, weighing in at about 60 kg (135 lbs.) and with no noticeable injuries or scars indicative of fights with other territorial males. It had recently fed, based on its protruding abdomen, and its paws (pads and claws) were in excellent condition. After we completed our examination, we administered the reversal chemical to revive the cat and moved to a safe distance to observe it, to ensure that it safely regained its senses and moved away from the capture site.

After the successful capture, we continued to collect location data on this individual for more than two years. Its territory was one of the largest in the Black Hills, covering an area that ranged from the south central (west of Custer along the Wyoming

FIGURE 3.1. The recapture of mountain lion M4. Note the rocky terrain that is typical along the South Dakota–Wyoming border of the Black Hills. M4 was initially pinned on a rock cliff by hounds. Steve Griffin and Jack Alexander were able to retrieve the hounds, allowing the cat to move to the adjacent hillside, where it was treed and recaptured. *Photo by John Kanta.*

border) to the central (Deerfield Lake and east to Route 385) region of the Black Hills. After the two years of obtaining locations on this male, we worried that the radio transmitter would fail (transmitters are generally guaranteed for about three years) or the collar material would break; either circumstance would end our contact with the lion. To prevent this from happening, we set out to recollar the male in early fall, just prior to the deer harvest season. Because the mountain lion was transmitting a signal, we flew over it to document its present location and then traveled by truck to a site close by, where were able to home in on the animal with a telemetry receiver and release hounds on its trail. We did so in a remote area of the Black Hills along the Wyoming border and followed it over two mountains before the hounds were able to tree it. We then immobilized the cat. At the time of the second capture, the male weighed about 78 kg (175 lbs.), and despite the large home range and two additional years of traversing a rather large portion of the region, there was no evidence (scars or wounds) indicative of territorial fights with other males. Again, we collected other biological data, reversed the immobilizing agents, and remained close to the animal for a period while it regained its composure.

We were pleased with the recapture of the male, but because of the adrenaline rush that accompanies wildlife and especially lion captures, we had lost track of the time. The temperature had been in the 60s, and because of the mild weather, I had gone off on the chase wearing just a sweatshirt (my graduate students were constantly amazed by my inability to bring sufficient clothing for captures of lions, deer, and other species). By the time we left the capture site, it was dark, the temperature had declined to below freezing, and the batteries in our GPS would not work properly. Luckily, there was enough snow on the ground for us to follow the tracks of other capture crew members who had left the site earlier. We met up with those crew members on one of the main roads in the Black Hills late that night, feeling cold and tired from the day's activities. For a few additional years after that capture, that male lion continued to help us understand how mountain lions were recolonizing the Black Hills.

General Information and Characteristics

The latin name of mountain lions, *Puma concolor,* means "cat of one color." Mountain lion fur varies in color from a tawny to a light cinnamon, with black on the back of the ears and the tip of the tail. Kittens are spotted, and the spots fade as they age. These faded spots, along with dark bars on the inner legs and the mottled appearance of the underfur, are used together with dental characteristics to age the species (fig. 3.2; Anderson and Lindzey 2000). Although cougars are currently classified as one subspecies (*P. concolor couguar*) in North America (Culver et al. 2000; Culver 2010), weights and body measurements vary geographically among populations (Maehr and Moore 1992). A general trend of increasing body size with increasing latitude applies to mountain lions (Kurten 1973; Iriarte et al. 1990).

Mountain lions exhibit sexual dimorphism, with males being about 60% larger than females. Adult males can be more than 2.5 m (more than 8 ft.) in length, including the tail, and weigh an average of 67 kg (150 lbs.). Adult females can be up to about 2 m (6.5 ft.) in length and weigh an average of 40 kg (90 lbs.). Generally, females tend to cycle from about 50 kg (112 lbs.) when pregnant to about 40 kg (90 lbs.) when lactating. Males, though, can continue to increase in weight up to about 78 kg in the Black Hills, if they can maintain a territory. In the Black Hills, adult females are approximately two-thirds the weight of adult males and approximately 17 cm shorter in total length (Thompson 2009; table 3.1).

Lions in the Black Hills have been aged to over 9 years for both males and females. Thompson (2009) measured 108 mountain lions (55 obtained during captures and 53 from mortality events) (table 3.1; fig. 3.3). Females had similar body measurements across age classes, with the exception of plantar pad lengths of front feet and neck circumference, both being larger in adult than in subadult females. Body measurements

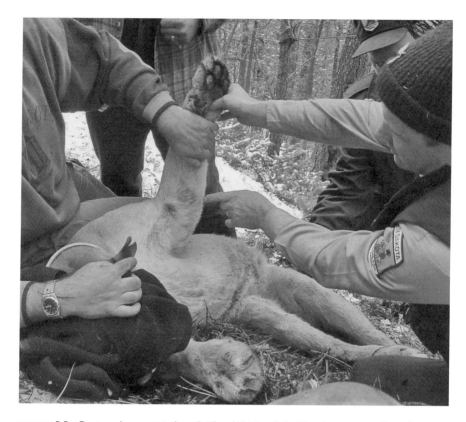

FIGURE 3.2. Captured mountain lion (M4) exhibiting faded barring on inner legs that was used, in addition to dental characteristics, to age the animal at 5–6 years. *Photo by D. Fecske.*

Table 3.1. Morphological characteristics (mean ± 1 SE) of mountain lions from the Black Hills of South Dakota and Wyoming, 1998–2007

	Adult male $n=25$	Adult female $n=32$	Subadult male $n=23$	Subadult female $n=27$
Age (yrs.)	5.51±0.57	4.14±0.24	1.92±0.10	1.66±0.05
Weight (kg)	63.16±1.99[ab]	41.68±1.01[b]	46.91±1.42[a]	36.79±1.15[b]
Total length (cm)	211.89±1.66[ab]	195.71±2.87[b]	205.89±2.81[a]	194.19±1.58[b]
Tail length (cm)	82.02±1.01	76.73±0.85	79.98±1.64	78.87±0.72
Chest circ.(cm)	76.16±1.08[ab]	65.55±0.85[b]	68.84±1.06[a]	63.37±1.11[b]
Neck circ. (cm)	41.45±0.67[ab]	35.36±0.41[b]	37.70±0.631[a]	34.07±0.696[b]
Front pad L (mm)	44.44±0.74[ab]	38.38±0.97[b]	40.35±0.74[a]	35.93±0.54[b]
Front pad W (mm)	62.54±0.84[ab]	52.38±0.76[b]	58.39±0.60[a]	52.30±0.54[b]
Rear pad L (mm)	42.08±0.78[ab]	36.06±0.60[b]	38.70±0.43[a]	34.63±0.50[b]
Rear pad W (mm)	53.72±0.75[ab]	45.44±0.56[b]	50.78±0.67[a]	46.30±0.64[b]

Source: Thompson 2009.
[a]Measurements differed between male age classes (P<0.05).
[b]Measurements differed between sexes (P<0.05).

FIGURE 3.3. Collecting biological data on a captured mountain lion. The young lion has begun to recover from immobilization. *Photo by South Dakota Department of Game, Fish and Parks.*

of male mountain lions differed by age class, and both adult and subadult males were larger than females; adult males averaged 63.2 kg (141 lbs.) and 211.9 cm (7 ft.) in length; subadult males averaged 46.9 kg (105 lbs.) and 205.9 cm (6.8 ft.) in total length.

Bobcats (*Lynx rufus*), the other resident wild cat in the Black Hills, weigh about 11 kg (25 lbs.), average 76 cm (2.5 ft.) in length, and have short tails (Mosby 2011). Despite the size and morphological differences between the two species, observations of bobcats, along with domestic cats, continue to be confused with those of mountain lions. Issues of depth perception probably account for such errors. The farther away an

FIGURE 3.4. Domestic "tabby" cat observed in a shelterbelt in North Dakota. The cat was reported as a mountain lion by the observer. *Photo from J. Ermer.*

animal is from the viewer, the larger it seems to be; thus tabby cats (fig. 3.4) that are observed in rather wild surroundings can be confused with mountain lions.

Mountain lions, especially adult males, are generally considered solitary animals. When lions are traveling together, it is usually an adult female and her kittens. In the Black Hills, kittens travel with their mother until 12 to 14 months of age (Thompson and Jenks 2010; Jansen and Jenks 2011); so the tracks of kittens can be similar in size to those of the adult female. Characteristics of mountain lion tracks include four toes, a three-lobed hind pad (similar to domestic cat tracks), and no claw prints because the claws are retractable (fig. 3.5). Despite the differences between canids (dogs and coyotes) and felids (bobcats and mountain lions), there have been numerous instances when canid tracks were misidentified as lion tracks. I suspect that this happens because observers generally look at the size of the track and assume that huge tracks are lion tracks (actually, most dispersing lions will weigh in the neighborhood of 80 pounds and have fairly small tracks). In addition, if you look at a set of tracks long enough, either in snow or in mud, there will be one track that resembles a lion track (the hind pad will seem to have three lobes and no claws will be distinguishable). However, if you look at the entire set of tracks, most will show claws and the classic hind pad of a canid. It therefore becomes important to look at as many tracks as possible when assessing the possibility that you are observing the tracks of a lion.

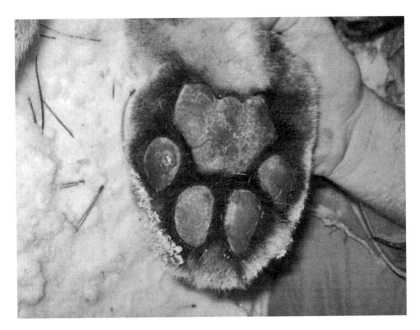

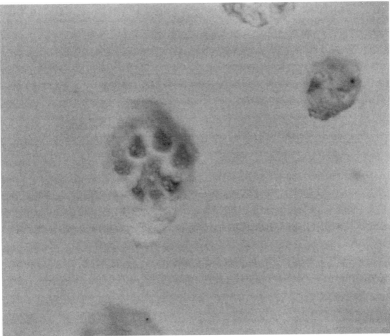

FIGURE 3.5. *Top,* a characteristic mountain lion paw, showing four toes, the three-lobed hind pad, and retracted claws. *Bottom,* a mountain lion print in snow. *Photos by Dan Thompson.*

Capturing Mountain Lions

During our studies of mountain lions in the Black Hills, we captured mountain lions more than 3 months old opportunistically, using trained dogs (Hornocker 1970), foot-hold snares (Logan et al. 1999), foot-hold traps (with offset jaws), wire-cage traps (Bauer et al. 2005), and the free-dart technique (capturing a free-ranging animal by delivering an immobilizing dart without prior restraint). We captured kittens less than 3 months old by hand (Logan and Sweanor 2001). We restrained young lions by placing them in a burlap sack (e.g., kittens less than 3 months old) or by injecting captured animals with a mixture of Telazol (tiletamine/zolazaline-hydrochloride) and xylazine-hydrochloride (Kreeger and Arnemo 2007); we always reconfirm drug dosages before setting out to capture wild animals.

Mountain lions were aged by tooth wear and pelage description (Anderson and Lindzey 2000), and animals more than 10 months old were fitted with either very high frequency or global positioning system radio transmitters (Telonics Inc., Mesa, AZ; North Star Science and Technology LLC, King George, VA; and Advanced Telemetry Systems, Isanti, MN). We counteracted xylazine-hydrochloride with yohimbine (Kreeger and Arnemo 2007), and, as with Lion M4, remained with sedated lions, observing them from a safe distance until they regained their ability to move away from capture sites.

When capturing a lion using hounds, we treed the cat (fig. 3.6) and erected a net around the tree prior to administering the immobilization drugs. After sufficiently immobilizing the animal, we climbed the tree (fig. 3.7), attached a rope to the lion, and lowered the animal to the ground. At times, the lion, upon being darted, would attempt to escape by jumping from the tree and into the net. Some would escape, while others that were partially immobilized would be supplementally drugged using a small dose of ketamine. The advantage of using ketamine as the supplemental drug is that the animal would not be doubly drugged with Telazol. Ketamine is metabolized independently from Telazol, so using this combination minimized the time the animal was immobilized.

Before immobilized mountain lions were given the reversal drug, yohimbine, we moved them to safe locations, relatively flat areas away from streams and rocky areas. At times, we also covered such lions with space blankets (or coats, especially during winter) prior to leaving the area, to ensure that drugged animals would maintain their core body temperatures while metabolizing the remaining drugs. When we were confident captured animals were in safe locations and were recovering from immobilization drugs, we would monitor the lions from a safe distance, using standard telemetry equipment.

At times we would look for the sign (tracks, scrapes) of mountain lions and set a bait site to attempt to capture animals (fig. 3.5). A potential site might be a saddle between two mountains that was crossed by lions or an area that was regularly marked

FIGURE 3.6. A mountain lion treed with the use of hounds in the Black Hills. *Photo courtesy of South Dakota Department of Game, Fish and Parks.*

with feces. Once a site for bait was established, we set a deer or other carcass at the site and monitored it with trail or video cameras (Thompson 2009; Jansen 2011). When we documented a lion using the bait, we would set traps (generally snares) around the carcass and set a transmitter on the snare, which would transmit a signal to a receiver if the trap was sprung. We then would approach the animal and immobilize it using standard darting procedures (figs. 3.8, 3.9). Because we were able to immobilize trapped animals quickly, we could assess the effects of the snare on the animals. Generally, any swelling associated with the snare would be gone within a few days

FIGURE 3.7. South Dakota Department of Game, Fish and Parks employee Steve Griffin, climbing a tree to lower an immobilized mountain lion for radio-collaring. *Photo by D. M. Fecske.*

after the trapping. Such quick recovery might have occurred because, when they were trapped, the animals were focused on the bait carcass and thus did not respond to the snare. These lions would generally be lying sternally recumbent when approached, allowing for quick darting. After darting an animal, we retreated and allowed it to remain calm while becoming immobilized.

On certain occasions, box or cage traps (fig. 3.10) were used to capture mountain lions. We used the design of Ken Logan and baited the trap with deer (Bauer et al. 2005). Once a lion was captured, we immobilized it using a jab pole. When it was safe to do so, we collected biological data, fitting the animal with a radio collar, and moved the animal to a safe location prior to reversing the immobilizing drugs. Box traps were generally used in areas close to housing developments, because it would be disruptive to run hounds through these areas; in addition, the tracks of domestic pets (dogs and

FIGURE 3.8. Graduate research assistant Brian Jansen (*center*) with the author (*right*), weighing a captured adult male mountain lion in Custer State Park. The mountain lion was captured using snares, with a deer carcass as bait. *Photo by students in a large mammal ecology and management class at South Dakota State University.*

cats) would potentially confuse the hounds during chases. As mentioned, all our capture methods were approved by the Institutional Animal Care and Use Committee at SDSU and followed recommendations of the American Society of Mammalogists (Gannon, Sikes, and Animal Care and Use Committee 2007; Sikes, Gannon, and Animal Care and Use Committee 2011, 2016).

FIGURE 3.9. Young mountain lions captured with the use of snares in the Black Hills. *Courtesy of B. Jansen.*

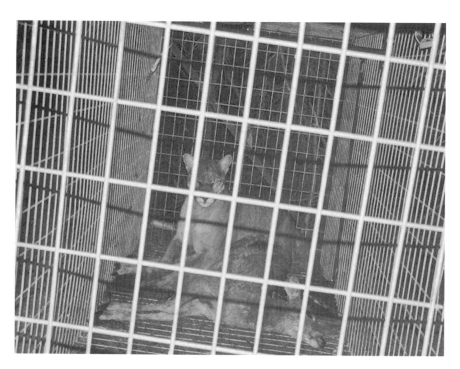

FIGURE 3.10. A mountain lion captured in the Black Hills using a box trap baited with a deer carcass. *Photo by Deanna Dawn.*

Literature Cited

Anderson, C. R., Jr., and F. G. Lindzey. 2000. A guide to estimating cougar age classes. Wyoming Cooperative Fish and Wildlife Research Unit, Laramie, WY.

Bauer, J. W., K. A. Logan, L. L. Sweaner, and W. M. Boyce. 2005. Scavenging behavior in Puma. Southwestern Naturalist 50:466–471.

Culver, M. 2010. Lessons and insights from evolution, taxonomy, and conservation genetics. In Cougar: Ecology and conservation, ed. M. Hornocker and S. Negri, 27–40. Chicago: University of Chicago Press.

Culver, M., W. E. Johnson, J. Pecon-Slattery, and S. J. O'Brien. 2000. Genomic ancestry of the American puma (*Puma concolor*). Journal of Heredity 9:186–197.

Gannon, W. L., R. S. Sikes, and the Animal Care and Use Committee of the American Society of Mammalogists. 2007. Guidelines of the American Society of Mammalogists for the use of wild mammals in research. Journal of Mammalogy 88:809–823.

Hornocker, M. G. 1970. An analysis of mountain lion predation upon mule deer and elk in the Idaho Primitive Area. Wildlife Monographs 21:1–39.

Iriarte, J. A., W. L. Franklin, W. E. Johnson, and K. H. Redford. 1990. Biogeographic variation of food habits and body size of the American puma. Oecologia 85:185–190.

Jansen, B. D. 2011. Anthropogenic factors affecting mountain lions in the Black Hills, South Dakota. PhD diss., SDSU.

Jansen, B. D., and J. A. Jenks. 2011. Body mass estimation in *Puma*. Wildlife Research 35:147–151.

Kreeger, T. J., and J. M. Arnemo. 2007. Handbook of wildlife chemical immobilization. 3rd ed. Terry J. Kreeger.

Kurten, B. 1973. Geographic variation in size in the puma (*Felis concolor*). Commentationes Biologicae 63:3–8.

Logan, K., and L. L. Sweanor. 2001. Desert puma: Evolutionary ecology and conservation of an enduring carnivore. Washington, DC: Island Press.

Logan, K. A., L. L. Sweanor, J. F. Smith, and M. G. Hornocker. 1999. Capturing pumas with foot-hold snares. Wildlife Society Bulletin 27:201–208.

Maehr, D. S., and C. T. Moore. 1992. Models of mass growth for 3 North American cougar populations. Journal of Wildlife Management 56:700–707.

Mosby, C. E. 2011. Habitat selection and population ecology of bobcats in South Dakota. Master's thesis, SDSU.

Sikes, R. S., W. L. Gannon, and Animal Care and Use Committee of the American Society of Mammalogists. 2011. Guidelines for the use of wild mammals in research. Journal of Mammalogy 92:235–253.

Sikes, R. S., W. L. Gannon, and Animal Care and Use Committee of the American Society of Mammalogists. 2016. Guidelines of the American Society of Mammalogists for the use of wild mammals in research and education. Journal of Mammalogy 97:663–688.

Thompson, D. J. 2009. Population demographics of cougars in the Black Hills: Survival, dispersal, morphometry, genetic structure, and associated interactions with density dependence. PhD diss., SDSU.

Thompson, D. J., and J. A. Jenks. 2010. Dispersal movements of subadult cougars from the Black Hills: The notions of range expansion and recolonization. Ecosphere 1:1–11.

CHAPTER 4

Population Dynamics of Mountain Lions

On one of my trips to the Black Hills, our plan was to fly over an area to locate the approximately 10 mountain lions we had successfully radio-collared. At the time, we were using conventional VHF (very high frequency) radio collars and thus had to be near the collar to receive the signal, via antenna and programmable receiver. Because mountain lions are constantly on the move (except when consuming a large kill or courting a female), we used aerial telemetry to find them on a weekly basis. The locations where we found the animals were used to assess survival, generate home ranges, and estimate the mountain lion population size.

On that trip, we met up with our pilot, known at the time as "Rodeo Bob," with whom we had contracted to conduct the flights, at the Rapid City Airport. I sat in the back of the plane so that my graduate student could use the telemetry receiver to find the radio-collared lions. The method involved listening for the signal (frequency) of a lion collar and essentially making an X flight over the estimated location while gauging the intensity of the signal (fig. 4.1). Where the flights crossed (the X) would be the estimated location of the animal. We also placed collars in random locations in the Black Hills to quantify the error of these estimates.

All was going well, in that we had found a few of the lions and had been in the air for about a half hour when I became a bit queasy from the circular pattern of the flight. I have always tended to become carsick; it happened on trips when I was young and on carnival rides at the local fair. Luckily, Bob had provided a "barf bag" to take care of my needs, and because of my gray appearance, he asked whether we should head back to the airport. I decided to put up with my distress, assuming that we would be airborne no more than another hour. I was most interested in ensuring that we collected the data on the lions, because of the cost of flights. Unfortunately for me, we remained in the air for another two hours. When we arrived back it the airport, I

FIGURE 4.1. Flying in search of radio-collared mountain lions above Interstate 90 along the northeast edge of the Black Hills. Interstate 90 lies between the Hogback and the Black Hills proper. Because of the topography and the thick vegetation (ponderosa pine, white spruce, and aspen) in the region, we rarely could actually see radio-collared mountain lions during aerial flights to estimate locations. *Photo by D. Fecske.*

struggled to remain upright as I walked from the plane to the terminal. Two hours later I opened my eyes and could stand and walk without feeling sick. I found out later that our pilot was aptly named "Rodeo Bob" because he intentionally tried to make his passengers sick. I only flew for lions one other time. I was seated in the front that time, and the pilot was told to be "kind" to me during the flight. I deplaned in much better shape.

When we began investigating mountain lions in the late 1990s, one of our original goals was to evaluate the amount of sign, such as tracks, as an index to population size within the southern region of the Black Hills. We hoped that these data would help to validate the estimate of 25–35 lions that was being used by South Dakota Department of Game, Fish and Parks at the time. Because mountain lions were purported to be at extremely low density, the capture of a sufficient sample to estimate population parameters such as survival and cause-specific mortality was thought unachievable. However, after a discussion of options with South Dakota Department of

Game, Fish and Parks officials and Fred Lindzey, assistant leader of the Wyoming Cooperative Fish and Wildlife Research Unit at the University of Wyoming, who had much experience with the species, it was decided that we would attempt to capture and radio-collar up to 10 individuals. Although still a small sample, at the time that number represented 20–40% of the estimated population of lions in the region, a significant proportion of the population.

We began our work in the southern Black Hills, because one of the trappers for the South Dakota Department of Game, Fish and Parks, Blair Waite, suggested that track observations in the vicinity of Custer State Park were highest in that region. My first graduate student on the project, Dorothy (Fecske) Wells, went to Wyoming to train with Chuck Anderson and his capture crew. Anderson was working on his PhD on mountain lions in the Snowy Mountains and had much experience capturing the species. Once the training was done, we hired a local houndsman who had experience with mountain lion chases in Montana. After our crew was assembled, we drove roads after fresh snow, first focusing on the area around Custer State Park and Spring Creek (north of the park), looking for tracks on which to release the hounds.

That first project netted capture of 12 adult lions and 2 kittens (Fecske 2003), but it also provided information that mountain lions might be greater in number than originally estimated and that they could be safely chased and treed using hounds within the Black Hills region. The great abundance of large ponderosa pines likely improved our success rate, because there were ample trees that lions could climb to evade the hounds. Also, the fact that captured lions ranged in age from just over 1 year to about 8 years indicated that lions likely had been occupying the Black Hills longer than expected and that there was a actively breeding population; nevertheless, the low population size and the secretive nature of the species had allowed them to evade notice by most locals and visitors to the Black Hills.

In addition to capturing mountain lions, we began necropsying mortalities documented in the Black Hills via our radio-collared sample and uncollared mortalities (carcasses reported to South Dakota Game, Fish and Parks that had been hit by vehicles or were found dead [Thompson 2009]). Deaths could and did result from natural causes: for example, lions fell through the ice of lakes or impoundments in spring, lions died from injuries sustained in fights between territorial males, one lion died from electrocution (Thompson and Jenks 2007), and another one died from smoke inhalation during the Jasper Fire (fig. 4.2; Fecske, Jenks, and Lindzey 2003). We were surprised at the number of deaths documented that were due to illegal shooting, trapping, and vehicles (human-caused mortality), which had not been documented in unhunted populations (Thompson, Jenks, and Fecske 2014; table 4.1). Although our sample size was low (12 lion carcasses), one other finding that was interesting was that most (more than 80%) of these lions showed evidence of interactions with porcupines (*Erethizon dorsatum*). Some carcasses had quills embedded in the front shoulders or around the facial area and neck of the carcass, while others had actually consumed

FIGURE 4.2. *Top,* a subadult male mountain lion that succumbed to injuries from an encounter with an adult territorial male in the southern Black Hills. *Bottom,* an adult female mountain lion that succumbed to smoke inhalation during the Jasper Fire. *Photos by D. M. Fecske.*

Table 4.1. *Characteristics of mountain lion mortalities in the Black Hills, South Dakota, 1996–2000*

N	Sex	Age	Fat[a]	Food[b]	Mortality	Location	Proximity[c]
1	M	8.5–9.5	L	PP	Shooting	Custer	South
1	M	2.5–3.5	L	NA	Vehicle	Spearfish	North
1	F	1.5–2.5	H	PP	Trap	Lake Pactola	Central
1	M	4–5 Months	M	NA	Vehicle	S. Hill City	South
1	F	3.5–4.5	NA	NA	Shooting	Pringle	South
1	F	3.5–4.5	H	PP	Capture	Custer	South
1	M	1.5–2.5	H	PP	Vehicle	Black Hawk	North
1	F	1.5–2.5	H	PP	Shooting	Deerfield Lake	Central
1	M	4.5–5.5	NA	NA	Shooting	Hot Springs	South
1	M	2.5–3.5	H	PP	Interaction	Custer	South
1	F	3.5–4.5	H	None	Fire	Jewel Cave	South
1	F	1.5–2.5	NA	NA	Shooting	Hot Springs	South

Source: Fecske, Jenks, and Lindzey 2003.
[a]Fat reserves were ranked as high (H), medium (M), or low (L) based on kidney fat.
[b]Catagories in the table are PP = porcupine, NA = not evaluated, None = GI tract empty.
[c]Proximity refers to region of the Black Hills.

the porcupines. Those that had consumed the prey had quills and paws in the stomach, suggesting that the cats had removed the paws after flipping the porcupine on its back. These cats also had quills throughout the remainder of the gastrointestinal tract, but those quills were quite rubbery, owing to the effects of stomach acid. The importance of this prey to mountain lions was further evidenced by a visual observation of a female carrying a porcupine carcass while traveling with her kittens; the observation occurred during helicopter surveys for mountain lions in the Black Hills (D. M. Fecske, SDSU, pers. commun.).

Other information obtained from mountain lion carcasses indicated that most (75%) of the animals came from the southern region of the Black Hills. This could be a function of our initial sampling of the population and could be related to the distribution of the human population, in that most visitors to the region spend time in Custer State Park and Wind Cave National Park. However, the information did support tracks found in the region and sightings when adjusted for human population density (see fig. 1.7). Also, most of these carcasses had high fat indices, which indicated that there was plenty of prey, in this case porcupines, as well as other prey within the parks, to support the low population. During one capture attempt in the southern portion of Custer State Park, I recall walking through snow 2–3 inches deep and kicking up bones of bison (*Bison bison*) and other species. Although these bones could have come from scavenged winter mortalities, that area of the park was ripe for capture of mountain lions during the early years of our studies.

One of the positive aspects of studying a long-lived species is that projects that extend over several years allow the accumulation of radio-collared individuals (Logan and Sweanor 2001). That was the case with our work. Although it can be difficult to maintain radio-collared mountain lions for long periods because of transmitter failure, interactions with other lions that affect the integrity of the collar material, deaths from natural and human-derived causes, and movement out of the Black Hills, some lions occupy rather predictable home ranges where they have access to adequate prey and cover habitat and are relatively far from human disturbance. These cats are likely to remain alive and well, and they may only need to be recollared every few years to remain active participants in research studies (see fig. 3.1). This was the case when we began evaluating survival, movements, and dispersal of mountain lions.

Studies had been conducted on these topics (survival, movements, and dispersal) previously (Ross and Jalkotzy 1992; Lindzey et al. 1994; Lambert et al. 2006), and most results indicated that lions generally had similar survival and home range sizes across regions, likely owing to the elaborate territorial system of the species (land tenure system) (Hornocker 1970; Seidensticker et al. 1973 [but see Pierce, Bleich, and Bowyer 2000; and Elbroch et al. 2016]). However, the Black Hills was somewhat isolated from other lion populations, and the habitats surrounding the Black Hills were almost moon-like (e.g., open prairies and few trees of a size that could be climbed or used for cover) to mountain lions raised in ponderosa pine habitat; such conditions might modify how this population used its environment. Questions that arose included these: Does increased population size affect territory and home range size? Does the relatively small size of the ecoregion force lions to disperse more often than in other populations?

Along with continuing to capture and affix transmitters to lions in our original focal area, the southern Black Hills, we began attempting to find and mark lions throughout the central and northern regions of the Black Hills. In addition, we transferred all carcasses that we accumulated to SDSU for necropsy. Using these carcasses, we documented the animals' condition, using kidney fat and prey consumed. We did, however, also notice that the number of carcasses was increasing with time (fig. 4.3). We hypothesized that the rate of increase in carcasses was in some way correlated with the increase in population size and thus could be used to estimate the intrinsic rate of increase of the population.

By means of a log transformation and regression analysis, I used the resulting slope of this relationship to estimate the intrinsic rate of increase of mountain lions, which was 0.33 and just a bit higher (but not significantly so, based on the standard error of the estimate) than the highest intrinsic rate of increase (0.28) presented for an increasing population of lions in New Mexico (Logan and Sweanor 2000). This rate translates to an innate ability for mountain lions to increase population size by up to 39% per year, and although this high rate might be specific to the Black Hills and due to the smorgasbord of prey available (from mountain goats to deer to small mammals and even birds, such as turkeys [*Meleagris gallopavo*]) to lions, it was still within the realm of rates considered reasonable for the species. Nevertheless, the estimate was just an index to the rate of increase for this population, because of the data used to

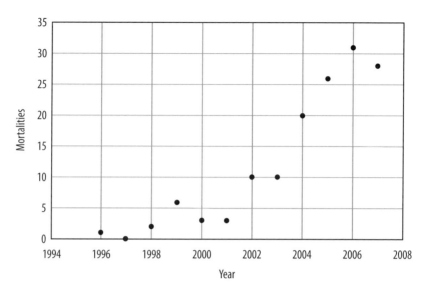

FIGURE 4.3. Relationship between mountain lion mortalities and time. These data were used to estimate rate of increase in the Black Hills population.

develop the value. We assumed that it was reasonable, however, because the region was semi-isolated from other known lion populations and was relatively small, there were few if any competitors, and the residents and visitors to the region were excited at the presence and possibility of observing the species and thus eager to report mortalities. In fact, the South Dakota Department of Game, Fish and Parks regularly received calls reporting sightings, vehicle-lion mortalities, and carcasses observed when visitors were recreating in the Black Hills. These conditions would enhance the likelihood that the relationship represented the pattern displayed by the population.

This piece of information also allowed us to model population size, beginning with the estimate generated by the South Dakota Department of Game, Fish and Parks of about 15 mountain lions in 1996 (fig. 4.4). Considering this potential for increase and the potential variability that could affect this rate of increase, including the initiation of a harvest, these data indicated that a somewhat protected mountain lion population (human-related mortality did occur during this time) could increase from about 15 individuals to more than 200 individuals in 8 to 10 years (1996–2006). Information in support of this increase included the increase in known mountain lion mortality, from just a few (1–2) to close to 30 per year over this time span. We were continually finding opportunities to capture new lions in the Black Hills, estimates of survival were greater than 80% for both adult males and females (table 4.2), and our preliminary estimates of population size using the captured lions in mark-recapture analyses supported high estimates of population size. Other studies that have documented survival of mountain lions over this time period have reported

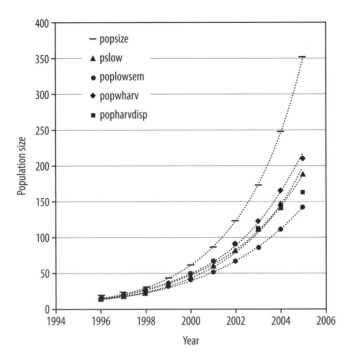

FIGURE 4.4. Initial population modeling based on starting population of 15 mountain lions and rate of increase of $r = 0.33$. Lines represent incorporation of the standard error about the estimated rate of increase and the addition of harvest for the Black Hills mountain lion population. Popsize = estimated population size with no mortality; pslow = population size, with mortality; poplowsem = pslow and one standard error lower rate of increase; popwharv = popsize with harvest; popharvdisp = popsize considering dispersal.

Table 4.2. *Survival rates for mountain lions tracked in the Black Hills from 1999 to 2005 pooled among years and separated by cohort*

Cohort	N	Survival (S ± SE)	Mean days tracked + SE
Adult Males	4	0.89 + 0.05	1569.5 + 481.5
Subadult Males	8	0.63 + 0.18	118.1 + 24.1
Adult Females	14	0.86 + 0.07	713.1 + 15.2
Subadult Females	5	1.00 + 0.00	274.3 + 37.9
Dependent Young[a]	15	0.67 + 0.11	155.1 + 34.3

Source: Thompson, Jenks, and Fecske 2014.
[a]Survival rates/days tracked for dependent young are reported from time of capture to date of death, date of censor, or date of independence (for individuals surviving to independence).

similar results (Clark et al. 2014). Also, the number of vehicle kills and removals increased, along with deaths from other causes (fig. 4.5), and 89% of resident males showed moderate to severe scarring from interactions with other males (fig 4.6) during this period of population increase and saturation (Thompson 2009). This last finding was much different from what we observed in our first captures, where adult males in the range of 3.5 to 6.5 years had virtually no injuries that would have been caused by interactions with other territorial males. When we combined all of our empirical information, we were confident that our estimates were reasonable for the species.

From about 1999 to 2005, we also captured 15 kittens from five unique litters (fig. 4.7), which averaged 3 kittens/litter initially, with a 5:1 male to female sex ratio (Thompson 2009). One kitten died from unknown causes, and one litter ($n = 4$) died owing to infanticide. The remaining litters and marked animals reached independence, resulting in a survival rate of 0.67, which was similar to what other studies had documented (Lindzey et al. 1988) but lower than for kittens in the Greater Yellowstone Northern Range (0.90 [Ruth et al. 2011]) and in west-central Montana (0.78 [Robinson et al. 2014]). Later estimates of kitten survival for work conducted after the start of the mountain lion harvest were based on the capture of 25 litters (Jansen 2011). Overall litter size was 3.0 kittens per litter, and the sex ratio was 5.33:1 (16 males, 3 females). Kitten survival averaged 0.62 ± 0.117 (Jansen 2011).

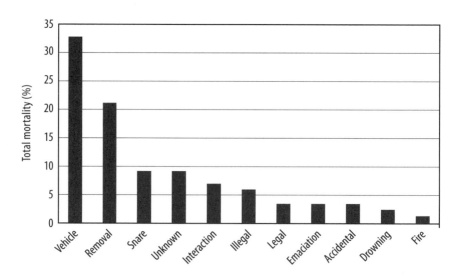

FIGURE 4.5. Mountain lions are vulnerable to many causes of mortality. Mortality events ($n = 85$) were documented for cougars in the Black Hills, 1998–2005. "Removal" refers to lions that were euthanized by the South Dakota Department of Game, Fish and Parks for safety (animals within towns and cities or those that caused deaths of domestic livestock or pets). "Accidental" mortalities were due to our capture activities. *Thompson 2009.*

FIGURE 4.6. Injuries received during a territorial dispute between two mountain lions. *Photo by Dan Thompson.*

Mountain lion kittens can be born in any month of the year, but there is a peak in the Black Hills (Jansen 2011; fig. 4.8). A birth pulse occurs from June to August (Jansen and Jenks 2012), but an interesting finding from our work was that the survival of kittens born within the birth pulse was similar to the survival of those born outside the pulse. In fact, one litter born in December survived to dispersal age.

Mean body mass for male ($n = 34$) and female ($n = 40$) kittens at 1 week of age was 1.2 kg (2.7 lbs.) and 1.1 kg (2.5 lbs.), respectively (Jansen and Jenks 2012). Between 2005 and 2009, we documented 31 litters that were born to females with whom we had recorded evidence of previous nursing. The birth timing period was narrower for females that had not previously nursed cubs ($n = 11$), and these females gave birth mostly in July ($n = 6$) as well as in June and August ($n = 3$). However, litters born after the peak of prey births (June–July) had heavier weights and dispersed at an older age, suggesting there was a benefit to these kittens because they were older and more experienced when they separated from their mothers (Jansen and Jenks 2012). Birth timing for females that had had previous litters ($n = 20$) were more dispersed throughout the year. This variation in birth timing likely occurs for older females because of kitten mortalities. After a female loses a litter, dependent upon prey availability, she

FIGURE 4.7. Mountain lion kitten radio-collared to estimate kitten survival. *Photo by B. Jansen.*

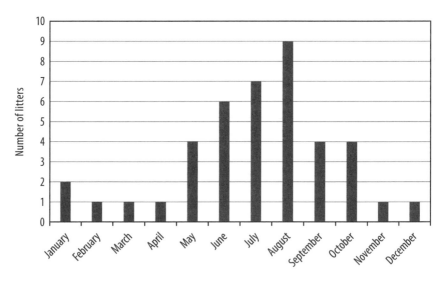

FIGURE 4.8. Frequency graph of litters of mountain lion kittens born in the Black Hills, by month. *Jansen and Jenks 2012.*

FIGURE 4.9. Kittens at den site in the Black Hills. *Photo by Dan Thompson.*

can become fertile again (physiologically in estrus) as soon as she acquires sufficient fat stores, then will breed, and a litter will be born about three months later (fig. 4.9).

We also investigated the use of placental scars for estimating litter size in mountain lions (fig. 4.10); the same technique has been used successfully for bobcat (McCord and Cardoza 1982) and lynx (*Lynx canadensis*) (Mowat, Boutin, and Slough 1996). However, our use of this technique came toward the end of our investigations

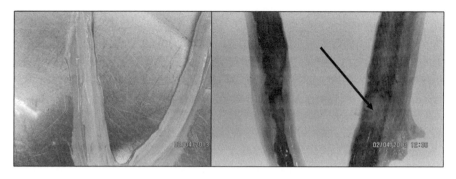

FIGURE 4.10. Comparison of 1.5-year-old (*left*) and 6.5-year-old (*right*) reproductive tracts of female mountain lions. The arrow points to a placental scar (previous attachment of placenta). *Photos by Brandi Felts.*

and thus provides but a snapshot rather than a pattern of how litter size might have changed temporally. Nevertheless, litter size determined from placental scars averaged 3.33 ± 0.48 kittens based on a sample of 30 females (B. Felts, SDSU, unpublished data). This average was similar to the 3.0 kittens per litter estimated by Jansen (2011). Furthermore, this sample indicated that first breeding of females occurred around 2 years of age and that litter size increased from about 1.0 to 5.0 kittens at 4–7 years of age.

Dispersal

During the period from 2003 to 2006, information we were collecting on this population suggested that mountain lions inhabiting the Black Hills were becoming "saturated" (i.e., exhibiting density-dependent effects [Thompson 2009]). Others who have studied large mammals, including mountain lions, have documented movements out of populations, especially when they approach carrying capacity. This process of dispersal (movement out of a population without return [Thompson and Jenks 2010]) is important to maintain genetic transfer among mountain lion populations via immigration and emigration (Sweanor, Logan, and Hornocker 2000; Culver et al. 2000; Anderson, Lindzey, and McDonald 2004) and to reduce strife among competitive individuals (Logan and Sweanor 2001; Maehr et al. 2002). Maehr and colleagues (2002) suggested that lions exhibit a "frustrated dispersal," which is predominately exhibited by males, but some females also move away from family members.

Generally, this dispersal movement is more than a home range away from family members, which could be problematic for lions in the Black Hills, considering the rather small width and length of the region and previously documented dispersal distances for the species (up to 224 km at the time [Maehr et al. 2002]). Yet, inexperience with the prairie habitat surrounding the Black Hills could represent a barrier to dispersal, as suggested by some of our first observations. There were cats that we believed were about to disperse that reentered the region rather than heading east across this unfamiliar terrain. For example, one of the first subadult males (the age class most likely to disperse) radio-collared was located during aerial surveillance in a cottonwood tree (*Populus deltoides*) along the Cheyenne River, suggesting it was about to move east out of the Black Hills and into the prairies of western South Dakota. However, later observations found the lion traversing the southern and western perimeter of the Black Hills prior to its departure from the region; the male finally moved to the Wyoming portion of the Black Hills and beyond to the west. In fact, these first observations indicated that at least some subadult males were likely leaving the Black Hills immediately after becoming independent of their mothers.

We were able to determine dispersal ages for 22 juvenile mountain lions (14 males, 8 females) (Jansen and Jenks 2012). Male mountain lions dispersed at a mean age of 14.7 months, and females dispersed at 15.3 months. Kittens born during spring and in the early and mid birth pulse (April 1–July 30) dispersed 4 months younger than kittens born in the last month of the birth pulse and during winter (August 1–January 31). Kittens that were born in early or mid pulse dispersed during the summer (June–August) or autumn (September–November) after attaining an age of 1 year; kittens born late in the birth pulse dispersed during winter (December–February) or spring (March–May) after becoming 1 year of age. Our findings provided support for the hypothesis that subadult mountain lions move out of the Black Hills at just about any time of the year; although there were times (September–October) when we expected that much of the dispersal activity occurred.

One problem we had with our attempts to document these dispersals was our use of traditional VHF radio-collar technology. These collars allowed us to pinpoint locations of lions from the air, but once they left the Black Hills, we were unable to determine where they moved or the direction of movement. Luckily, other states and provinces in the region were happy to provide information on lions traveling through their areas. Our first experience with this system of information transfer was a call received from the Oklahoma Department of Wildlife Conservation. They called to confirm that a lion killed by a train outside of Redrock, Oklahoma, was an animal collared in the Black Hills (Thompson and Jenks 2005). Because of the distance from the Black Hills, I asked how they had confirmed that it was an animal from our studies, and the response was that they had contacted the telemetry company (Telonics Inc.) that had produced the collar, and the company confirmed that the serial number on the radio collar was sold to SDSU. This cat had dispersed 660 miles (1,067 km straight-line distance) from the last location we had obtained on it in the Wyoming portion of the Black Hills to the location of its death in Oklahoma.

Since that first confirmed dispersal, we were able to document lions moving from the Black Hills to North Dakota and northern Minnesota, to Nebraska along the Niobrara River, to northwestern South Dakota, to the Yellowstone River in central

Montana, to the Judith Mountains in Montana (Thompson and Jenks 2010; fig. 4.11), and more recently, to the Bighorn Mountains in Wyoming, to Saskatoon, Saskatchewan, and to Lake Sakakawea in North Dakota. Because these movements were so extensive and involved so many surrounding states and provinces, just about every lion documented east of the Black Hills was linked to our work or to the Black Hills region. For example, a lion that was believed to have moved through Wisconsin and was subsequently killed in downtown Chicago was linked to the Black Hills. Although it was not collared, the cat was linked genetically to the Black Hills, based on work we were conducting at that time (see chapter 7; Thompson 2009; Juarez et al. 2016). A lion killed by a vehicle in Connecticut also was linked genetically to the Black Hills region, using the same methods (Hawley et al. 2016). At the time, Missouri was receiving reports of mountain lions, and although undocumented, these reports also were anecdotally linked to the Black Hills. Despite these close ties to the Black Hills, other mountain lion studies were being conducted in Colorado and Montana, and mountain lions collared in Colorado were traveling through western Nebraska. These lions likely traveled farther east and could be responsible for some of the dispersers establishing east of the Black Hills region.

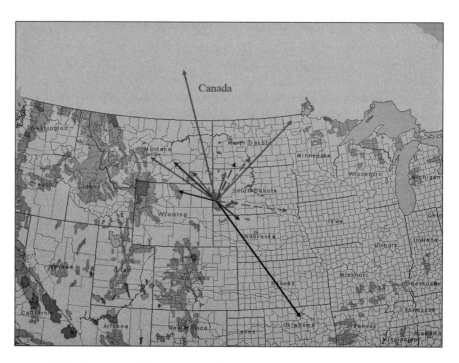

FIGURE 4.11. Documented dispersals of radio-collared mountain lions from the Black Hills region. *Adapted from Thompson and Jenks 2010.*

FIGURE 4.12. Approximate travel of a mountain lion from the Black Hills to Connecticut, where the animal was killed by a vehicle. Numbers represent locations were DNA evidence was collected or the animal was documented via camera; number 10 is the location where the mountain lion died. *Hawley et al. 2016.*

Based on our capture of 14 subadult male and 10 subadult female mountain lions, dispersal averaged 275 km (13–1067 km) for males and 48 km (12–99 km) for females (Thompson and Jenks 2005, 2010). These estimates do not include the Connecticut mountain lion that dispersed from 2,450 to more than 2,700 km (1,531 to more than 1,687 miles) (Hawley et al. 2016; fig. 4.12). All of the subadult males and 50% of the subadult females dispersed from their natal range (although, because of mortality, not all males left the Black Hills). We attributed the rather high dispersal rate of females to the saturated condition of the population, whereas the high dispersal rate of males was similar to findings in other studies of this behavior in mountain lions.

Despite the problems associated with using conventional VHF radio technology in documenting dispersal of mountain lions, we were successful, and along with the straight-line distances traveled, the data collected allowed calculation of daily rates of movement as an index to how fast mountain lions traveled across the landscape. For example, for four subadult males, the time between locations ranged from 7 to 267 days, while the distance traveled ranged from 76 km to 1,067 km. Our rough estimate of daily rates of travel averaged 8 km/day (5 miles/day) and ranged from 4 to 13 km per day (2.5 to 8 miles per day) (table 4.3). These findings were minimum values

Mountain lion	Days between locations	Total distance traveled	Travel rate
M16	267	1,067 km	4.0 km/day
M17	21	100.5 km	4.8 km/day
M19	7	75.6 km	10.8 km/day
M21	14	183.6 km	13.1 km/day
Average	77.25	356.7 km	8.2 km/day

Source: D. J. Thompson.

because they are based on straight-line distances, which are not characteristic of how mountain lions travel. This high rate of travel and the type of terrain traversed make it difficult to keep up with dispersing lions. These issues also help to explain why people who describe mountain lion sightings to wildlife professionals might not have their observations verified; those lions could be 30 km away from the sighting location by the time officials arrive at the site. Furthermore, lions are light-footed, and these younger subadult lions that weight about 36 kg might not leave identifiable tracks at the location of the sighting.

Home Ranges

When we began studying mountain lions, the territory size of male cats seemed enormous, compared to the findings of other studies. For example, the average territory size for three males and three females was 797 km^2 (308 miles^2) and 159 km^2 (61 miles^2) (Fecske 2003). This average home range size for males and females represented about 9% and 2% of the entire Black Hills region and would imply that there was enough area in the Black Hills to support about 11 territorial males and about 50 females. For males, at least, these large territories likely were a function of the low population size, and during our activities to locate radio-collared males, we noted movements across territorial borders. For example, Route 385 formed the border between the territories of two adult males; females were located on each side of the highway. Males had been observed crossing the highway, and when captured they seemed to be in excellent condition relative to battle scars, suggesting that home ranges were so large that the probability that the two males encountered each other was low, even when visiting and/or breeding with females within each other's territory. Later home ranges, estimated when the population was approaching saturation (i.e., the capacity of the Black Hills to support mountain lions), ranged in size from 251 to 2,314 km^2 for males and from 55 to 623 km^2 for females (table 4.4; fig. 4.13). The home range size differential was 3.7 to 4.6 times as large for males as for females; this difference was a bit larger than that documented for mountain lions in northwestern Wyoming (1.4–1.9 [Lendrum et al. 2014]).

Table 4.4. *Variability in annual home range estimates (km; 90% adaptive kernel)*
of 13 adult mountain lions in the Black Hills, South Dakota

Lion	Year	Home Range	N	Lion	Year	Home Range	N
M1	1999	854.1	52	M15	2003	623.4	32
M1	2000	1219.7	37	F6	2000	395.0	48
M1	2001	1329.7	41	F6	2001	127.1	48
M1	2002	853.8	66	F6	2002	185.7	77
M2	1999	462.1	46	F8	2000	206.7	27
M2	2000	900.0	48	F9	2000	187.2	44
M2	2001	847.5	34	F9	2001	74.8	47
M3	1999	297.7	25	F9	2002	298.2	69
M4	1999	251.4	34	F10	2000	429.3	40
M4	2000	689.1	42	F10	2001	241.0	45
M4	2001	729.3	39	F10	2002	260.5	32
M4	2002	786.4	58	F14	2003	117.7	38
M12	2000	2314.7	36	F18	2003	108.7	23
M12	2002	1376.8	51	F20	2003	55.6	34

Source: D. J. Thompson.
Note: M = male; F = female.

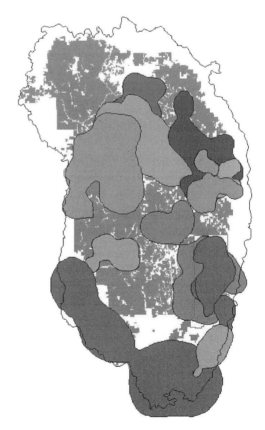

FIGURE 4.13. Early home range
contours of mountain lions captured
and radio-collared in the Black Hills.
Courtesy of Dan Thompson.

These early, large home ranges were in part a function of the area where we began our work; however, initial estimates seemed to indicate that the home range sizes were declining with increased population size. Jansen (2011) compared home ranges of male and female lions captured in various regions of the Black Hills. He found that both males (804 km²) and females (198 km²) in the southwestern region of the Black Hills had the largest home ranges, whereas males (315 km²) in the northwest and southeast and females (66 km²) located in the northeast had the smallest home ranges. Although he did not speculate on the potential reasons for these larger home ranges in the southwestern region of the Black Hills, it is possible that the distribution of the prey base might be involved.

Early in our work with mountain lions, we also were investigating deer ecology in the southern Black Hills. As mentioned, we did not document mortalities associated with lions in the north and central regions of the Black Hills; however, we did document mortalities of deer, both white-tailed and mule deer, caused by lions in the southern region. At the time, we had both white-tailed and mule deer radio-collared, and data collected on mortalities suggested that coyotes (*Canis latrans*) were the primary predator of white-tailed deer and that lions were focusing on mule deer. Because mule deer occupied the brushy-scrub fringe of the Black Hills that also was characterized by mountain mahogany (*Cercocarpus montanus*) and mountain juniper (*Juniperus scopulorum*) (Simpson 2015), this habitat could have benefited lions when attempting capture of mule deer. White-tailed deer were primarily found in open ponderosa pine habitats, where it would be more difficult for lions to pursue and capture them. Therefore, lions in the southwest might have had a more diverse prey base they were attempting to prey upon; if so, that fact might have resulted in a need for larger areas (Jansen 2011) to support successful capture of prey.

As with variability in home range size, overlap in territories between adjacent male mountain lions varied considerably (table 4.5). This variation (from 0 to 52%) was likely due to a number of factors, including the distribution of female home ranges within male territories, the number of females, the ages of adjacent males, the topographical characteristics of home ranges, and the presence of transient males.

Table 4.5.   Home range overlap (%) between 3 male mountain lions

	M1 M2	M1 M2	M1 M4	M1 M4	Total % overlap for M1	
Year	90%	60%	90%	60%	90%	60%
1999	5.1	0.0	12.8	0.0	17.9	0.0
2000	15.3	2.3	36.7	20.2	52.0	22.0
2001	16.0	12.0	15.0	0.0	28.0	12.0
Average	12.1	4.8	21.5	6.7	32.6	11.3

Source: D. J. Thompson.
Note: Overlap zones were identified using 90% and 60% AK home ranges.

Population Compensation and Harvest

Most studied prey (e.g., deer) populations exhibit compensation when approaching carrying capacity. This response is characterized by increased mortality owing to decreased availability and quality of forage, which worsens the nutritional condition of individuals in the population, reduces reproduction, and results in the population establishing an equilibrium at some ecological carrying capacity. Based on the population trajectory and increased starvation that we documented, which coincided with a perceived population saturation of mountain lions, we assumed that this population, rather than exhibiting a classic predator-prey type of response, was responding more to food availability and thus in a manner similar to that of a classic prey population. If such a response was occurring, then reducing population size postsaturation would result in a decrease in mortality and a corresponding increase in survival. This standard response would be due to the subsequent increase in prey availability resulting from a decrease in predators, in this case, the mountain lions themselves.

Preliminary information collected on this population supported this hypothesis. For example, vehicle mortalities accounted for 33% of mountain lion mortalities documented prior to the initiation of harvest (2005) in the Black Hills (Fecske, Thompson, and Jenks 2011). During the three years following the harvest, vehicle mortality declined from 22.5% in 2005, to 16.1% in 2006, to 8.9% in 2007, while the harvest rate on this population increased to 14% (based on harvest of radio-collared lions) and total mortality was relatively stable at 56 ± 6 mountain lions (Fecske, Thompson, and Jenks 2011). The nutritional condition of the lions also improved in

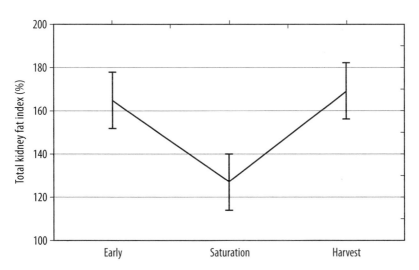

FIGURE 4.14. Change in estimated total kidney fat index across three periods: early, saturation, and harvest.

years when harvest was allowed (fig. 4.14), thus providing circumstantial evidence of reduced density dependence with increased food availability. However, additional harvest provided more detailed information on mortality and indicated an increase in mortality after the initiation of harvest; thus, any compensation between natural and harvest mortality was minimal and difficult to quantify (i.e., once the population had begun to decrease, total mortality increased, indicating that harvest mortality was additive [added to natural mortality] rather than being compensated by natural mortality). More recent studies of this phenomenon have provided support for both types of mortality: additive and partially compensatory (Wolfe et al. 2015). The size of the study areas evaluated by Wolf and colleagues (2015) were 11% and 15% of the entire Black Hills region, which might suggest that there could be portions of the Black Hills where mountain lion mortality is partially compensatory; thus, the scale evaluated by Fecske, Thompson, and Jenks (2011) may have masked local effects.

Population Modeling

From our long-term collection of information on population size, mortalities, litter size, survival, and sex ratio, as well as reference to published information on population dynamics of other mountain lion populations, we were able to construct a population model for Black Hills lions. We had used population reconstruction to model the population to carrying capacity (K), and at that time we had generated estimates of population size via mark-recapture models (Jansen 2011), we merged three approaches to generate a population curve through time (fig. 4.15). We used population reconstruction and rate of growth to generate the curve to 2007, mark-recapture estimates for 2008–2009, and reproductive information and mortality, including harvest, to model the population to 2013.

For the mark-recapture estimates, we attempted to maintain a sample of 50 radio-collared lions, and we focused our efforts on adults, to improve the chances that the animals would remain in the Black Hills during the harvest season. Variation in the harvest of radio-collared mountain lions and the low harvest of radio-collared males, in part because a lower number of adult males than females were radio-collared, resulted in reasonable estimates for females but not for other age and sex classes of this population.

Our model is a representation of how this population likely changed through time, but like any model, it suffers from a number of unknowns. For example, we knew that a large proportion of subadult males were leaving the region (emigrating) but had virtually no information to generate rates of immigration for the model; thus, we assumed that these rates were equivalent to one another. Nevertheless, the semi-isolated nature of the Black Hills and long-term study of the population helped to limit external influences on this population. One other issue that should be addressed is the lack of variation in our annual estimates. As mentioned, the model is a representation of

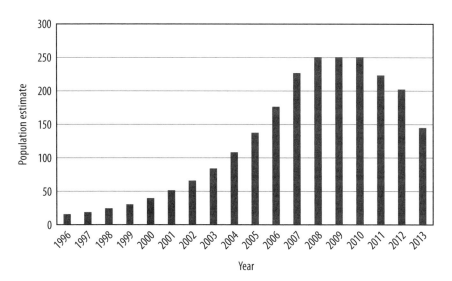

FIGURE 4.15. Refined population model for mountain lions in the Black Hills based on preliminary modeling (see fig. 4.3) and modified based on mark-recapture results and harvest reduction. *J. A. Jenks, unpublished data; Juarez et al. 2016.*

the change in population through time. We believe that pattern is reasonable, but the absolute estimates are, at best, conservative, because of the lack of information on mountain lions immigrating to the Black Hills and the fact that the information collected during these studies was not uniformly distributed over the entire region. As a consequence, there could have been additional mountain lions that would increase true estimates of population size above these estimates. From a management perspective, such conservative estimates might be considered beneficial, because the potential for overharvest would be minimized. Alternatively, from a strict management perspective, where reduction in the mountain lion population is the focus, such an approach might be considered preservationist.

Literature Cited

Anderson, C. R., Jr., F. G. Lindzey, and D. B. McDonald. 2004. Genetic structure of cougar populations across the Wyoming Basin: Metapopulation or megapopulation. Journal of Mammalogy 85:1207–1214.

Clark, D. A., B. K. Johnson, D. H. Jackson, M. Henjum, S. L. Findholt, J. J. Akenson, and R. G. Anthony. 2014. Survival rates of cougars in Oregon from 1989 to 2011: A retrospective analysis. Journal of Wildlife Management 78:779–790.

Culver, M., W. E. Johnson, J. Pecon-Slattery, and S. J. O'Brien. 2000. Genomic ancestry of the American puma (*Puma concolor*). Journal of Heredity 9:186–197.

Elbroch, L. M., P. E. Lendrum, H. Quigley, and A. Caragiulo. 2016. Spatial overlap in a solitary carnivore: Support for the land tenure, kinship, or resource dispersion hypotheses? Journal of Animal Ecology 85:487–496.

Fecske, D. M. 2003. Distribution and abundance of American martens and cougars in the Black Hills of South Dakota and Wyoming. PhD. diss., SDSU.

Fecske, D. M., J. A. Jenks, and F. G. Lindzey. 2003. Characteristics of mountain lion mortalities in the Black Hills, South Dakota. In Proceedings of the Sixth Mountain Lion Workshop, 25–29. Austin: Texas Parks and Wildlife Department.

Fecske, D. M., D. J. Thompson, and J. A. Jenks. 2011. Natural history and ecology of cougars. In Managing cougars in North America, ed. J. A. Jenks, 15–40. Logan, UT: Berryman Institute Press; Western Association of Fish and Wildlife Agencies.

Hawley, J. E., P. W. Rego, A. P. Wydeven, M. K. Schwartz, T. C. Viner, R. Kays, K. L. Pilgrim, and J. A. Jenks. 2016. Long-distance dispersal of a subadult male cougar from South Dakota to Connecticut documented with DNA evidence. Journal of Mammalogy 96:1435–1440.

Hornocker, M. G. 1970. An analysis of mountain lion predation upon mule deer and elk in the Idaho Primitive Area. Wildlife Monographs 21:1–39.

Jansen, B. D. 2011. Anthropogenic factors affecting mountain lions in the Black Hills, South Dakota. PhD diss., SDSU.

Jansen, B. D., and J. A. Jenks. 2012. Birth timing for mountain lions (*Puma concolor*): Testing the prey availability hypothesis. Plos One 7(9):e44625.

Juarez, R. L., M. K. Schwartz, K. L. Pilgrim, D. J. Thompson, S. A. Tucker, and J. A. Jenks. 2016. Assessing temporal genetic variation in a cougar population: Influences of harvest and neighboring populations. Conservation Genetics 17:379–388.

Lambert, C. M. S., R. B. Wielgus, H. S. Robinson, D. D. Katnik, H. S. Cruickshank, R. Clarke, and J. Almack. 2006. Cougar population dynamics and viability in the Pacific Northwest. Journal of Wildlife Management 70:246–254.

Lendrum, P. E., L. M. Elbroch, H. Quigley, D. J. Thompson, M. Jimenez, and D. Craighead. 2014. Home range characteristics of a subordinate predator: Selection for refugia or hunt opportunity. Journal of Zoology 294:59–67.

Lindzey, F. G., B. B. Ackerman, D. Barnhurst, and T. P. Hemker. 1988. Survival rates of mountain lions in southern Utah. Journal of Wildlife Management 52:664–667.

Lindzey, F. G., W. D. Van Sickle, B. B. Ackerman, D. Barnhurst, T. P. Hemker, and S. P. Laing. 1994. Cougar populations dynamics in southern Utah. Journal of Wildlife Management 58:619–624.

Logan, K. A., and L. L. Sweanor. 2000. Puma. In Ecology and management of large mammals in North America, ed. S. Demarias and P. R. Krausman, 347–377. Upper Saddle River, NJ: Prentice-Hall.

Logan, K., and L. L. Sweanor. 2001. Desert puma: Evolutionary ecology and conservation of an enduring carnivore. Washington, DC: Island Press.

Maehr, D. S., E. D. Land, D. B. Shindle, O. L. Bass, and T. S. Hoctor. 2002. Florida panther dispersal and conservation. Biological Conservation 106:187–197.

McCord, C. M., and J. E. Cardoza. 1982. Bobcat and lynx. In Wild mammals of North America: Biology, management, and economics, ed. J. A. Chapman and G. A. Feldhamer, 728–766. Baltimore: Johns Hopkins University Press.

Mowat, G., S. Boutin, and B. G. Slough. 1996. Using placental scar counts to estimate litter size and pregnancy rate in lynx. Journal of Wildlife Management 60:430–440.

Pierce, B. M., V. C. Bleich, and R. T. Bowyer. 2000. Social organization of mountain lions: Does a land tenure system regulate population size? Ecology 81:1533–1543.

Robinson, H. S., R. Desimone, C. Hartway, J. A. Gude, M. J. Thompson, M. S. Mitchell, and M. Hebblewhite. 2014. A test of the compensatory mortality hypothesis in mountain lions: A management experiment in west-central Montana. Journal of Wildlife Management 78:791–807.

Ross, P. I., and M. G. Jalkotzy. 1992. Characteristics of a hunted population of cougars in southwestern Alberta. Journal of Wildlife Management 56:417–426.

Ruth, T. K., M. A. Haroldson, K. M. Murphy, P. C. Buotte, M. G. Hornocker, and H. B. Quigley. 2011. Cougar survival and source-sink structure on Greater Yellowstone's Northern Range. Journal of Wildlife Management 75:1381–1398.

Seidensticker, J. C., IV, M. G. Hornocker, W. V. Wiles, and J. P. Messick. 1973. Mountain lion social organization in the Idaho Primitive Area. Wildlife Monographs 35:1–60.

Simpson, B. D. 2015. Population ecology of Rocky Mountain elk in the Black Hills, South Dakota and Wyoming. Master's thesis, SDSU.

Sweanor, L. L., K. A. Logan, and M. Hornocker. 2000. Cougar dispersal patterns, meta-population dynamics, and conservation. Conservation Biology 14:798–808.

Thompson, D. J. 2009. Population demographics of cougars in the Black Hills: Survival, dispersal, morphometry, genetic structure, and associated interactions with density dependence. PhD diss., SDSU.

Thompson, D. J., and J. A. Jenks. 2005. Long-distance dispersal by a subadult male cougar (*Puma concolor*) from the Black Hills, South Dakota. Journal of Wildlife Management 69:818–820.

Thompson, D. J., and J. A. Jenks. 2007. Cougar mortality attributed to electrocution from power lines in South Dakota. Prairie Naturalist 39:191–193.

Thompson, D. J., and J. A. Jenks. 2010. Dispersal movements of subadult cougars from the Black Hills: The notions of range expansion and recolonization. Ecosphere 1:1–11.

Thompson, D. J., J. A. Jenks, and D. M. Fecske. 2014. Prevalence of human-caused mortality in an unhunted cougar population and potential impacts to management. Wildlife Society Bulletin 38:341–347.

Wolfe, M. L., D. N. Koons, D. C. Stoner, P. Terletzky, E. M. Gese, D. M. Choate, and L. M. Aubry. 2015. Is anthropogenic cougar mortality compensated by changes in natural mortality in Utah? Insight from long-term studies. Biological Conservation 182:187–196.

CHAPTER 5

Disease Ecology of Mountain Lions

I have had a scientific curiosity concerning wildlife diseases since my experiences as a graduate student studying deer nutrition. In fact, one of my first publications was about a fawn that exhibited skull deformities and was assumed to be abandoned by its mother because it could not suckle (Jenks, Leslie, and Gibbs 1986). I realized that most wildlife studies at the time (especially in northern states) considered diseases and disease-related mortality to have minimal impacts on wildlife populations, and that those who studied movements, survival, or population dynamics focused more on human or predator impacts, rather than diseases, no matter the focal species (deer, elk, mountain sheep) under study. But I still wondered about interactions among diseases and nutritional condition, interactions among diseases and the probability of selection of prey by predators, and the full extent of the effect of disease-related mortality on the maintenance of wildlife populations.

Not long after I arrived in South Dakota, a new disease, chronic wasting disease, was documented in elk in Wind Cave National Park. It was thought that captive elk had been transferred from an area in Colorado where the disease was prevalent to a private ranch just south of Wind Cave National Park. Those elk had then interacted with free-ranging elk within the park or elk that moved out of the park via the perimeter fence, which was about 1 m (3 ft.) in height along the southwest border of the park (Bauman, Jenks, and Roddy 1999). Elk that crossed the fence line moved into the adjacent Black Hills National Forest and into close proximity with the captive elk. Despite the high probability of transfer of the disease from captive to wild elk, there was little known about the status of the disease in South Dakota. Some within the captive animal industry questioned whether the disease had always been in the Black Hills region as well as other regions within the state.

Because there was no information available to answer this question, we began a study funded through the South Dakota Department of Game, Fish and Parks to collect brain tissue from hunter-harvested deer and elk to test for the presence of the disease (Jacques et al. 2003). Heads of legally harvested deer and elk from throughout South Dakota were transported to the laboratories of the South Dakota Department of Game, Fish and Parks in Rapid City and the SDSU laboratories in Brookings, and brain tissue samples were collected and sent to the Diagnostic Laboratory at Iowa State University to test for the disease. This study provided evidence that the disease was present in free-ranging deer but only in the same area as where those captive elk were located (Jacques et al. 2003) and in very low prevalence; thus, statements associating the disease with captive elk were supported. These findings also established the need to collect information on other diseases potentially affecting large mammals in the region.

One study that included a focus on disease dealt principally with the effects of fire on the nutritional condition of white-tailed and mule deer in the southern Black Hills (Zimmerman 2004). The student who carried out the study, Teresa (Zimmerman) Frink, was intensely interested in diseases and, because of this interest, dedicated part of her thesis to this topic. We also knew that some deer in this region were dying because of predation by mountain lions. The deer study involved the removal of adult females for necropsy and sample collection. During these collections we obtained information on external and internal parasites as well as exposure to diseases documented by blood serum samples (Zimmerman 2004). One interesting finding was the presence of the parasite *Taenia omissa* in mule deer (fig. 5.1). The parasite requires an intermediate host and generally passes between predators and prey species. Predators carry the parasite, which is dispersed within the environment via feces. Prey, such as mule deer, ingest the parasite when consuming vegetation. *Taenia omissa* is considered a core species within the distribution of mountain lions (Waid and Pence 1988), and the presence of the parasite within mule deer from 2002 to 2003 provided additional support for an active predator-prey relationship between the two species early in our studies of mountain lions. This same relationship was documented with the recent presence of mountain lions in Manitoba (Dare and Watkins 2012).

In the late 1990s, when we began our work, those first mountain lions captured appeared healthy, based on external coat, teeth characteristics, and body weights. This relatively good health was supported by necropsy of mortalities, which included collared as well as noncollared cats. Also, as mentioned, males had few scars that would indicate intrasexual battles over territories as had been documented in other studies (Maehr 1997; Logan and Sweanor 2001). However, as the population increased in size from 1999 to 2005, we began seeing potential issues with the health of mountain lions in the Black Hills.

One of the most notable changes was first observed in the southern Black Hills and involved what we called cloudy-eye syndrome (fig. 5.2), a corneal opacity caused by

FIGURE 5.1. A *Taenia* cyst collected from deer in the southern Black Hills in the early years of the 2000s. *Taenia omissa* is a common parasite of mountain lions and has been documented in mountain lions throughout their range. *Photo by Teresa (Frink) Zimmerman.*

neutrophil infiltration of the corneal tissues (Jansen 2011). We began capturing lions and received reports of cats that appeared to be going blind (D. J. Thompson, pers. commun.). One observation was of a female in Custer State Park who was commonly found at a "bone pile," suggesting that the cat was unable to make kills and had resorted to scavenging. A second observation was of a relatively young female (about 3.5 years old) with kittens that was occupying an area close to a campground in the central Black Hills. Despite concern that we might end up documenting a first attack on a human in the Black Hills—a partially blind lion attempting to capture nontypical prey (i.e., humans)—the decision was made not to euthanize but to keep watch on this individual. Fortunately, the cat did not attack anyone, and her eyes seemed to eventually clear to some extent (fig. 5.2, bottom).

One of my graduate students, Dan Thompson, transported the carcass of one individual that had this cloudy-eye condition to the Animal Research Diagnostic Laboratory at SDSU in an attempt to document the pathogen involved with the disease. Preliminary diagnosis suggested that the pathogen was *Chlamydia,* and unfortunately, the local media released this preliminary information. As a result, *USA Today* published a story stating that mountain lions in the Black Hills were infected with

FIGURE 5.2. Change in eye characteristics of a mountain lion with cloudy eye syndrome. *Top,* a healthy mountain lion; *middle,* a mountain lion with cloudy cornea; *bottom,* a mountain lion that has recovered from the disease. Note: those lions that recovered continued to display partial cloudiness of the cornea. *Photos by Brian Jansen.*

sexually transmitted diseases. Additional analyses, however, did not support *Chlamydia* as the primary pathogen involved with this syndrome.

We continued to attempt to document the pathogen by swabbing the eyes of captured lions and submitting samples to a laboratory at Colorado State University (Jansen 2011). At the time when the disease was most prevalent, I estimated that 5–10% of the mountain lions in the Black Hills suffered from it and suspected that it might have been transferred to mountain lions when they were killing and consuming domestic cats, actions that became more common when the population was believed to be saturated (e.g., 2005). Unfortunately, we were only able to conclude that these lions suffered from a "nonspecific uveitis," which disappeared from the population as quickly as it had arisen, once the lion population had begun to decline owing to hunter harvest. Although unconfirmed, the disease could be related to the relatively high prevalence of feline calicivirus (about 5% based on ocular swab samples [table 5.1]; about 27% based on blood sera samples [table 5.2])., documented concurrent with the uveitis in the lion population (Jansen 2011). However, lions and other cats affected by feline calicivirus most commonly display influenza-like symptoms and not the cloudy-eye syndrome we had documented. Other potential pathogens (i.e., Herpes, *Mycoplasma*), in addition to *Chlamydia*, also were extremely rare in these lions (table 5.1).

Because of the methods used to collect data for our projects, we rarely came in direct contact with research animals for extended periods of time (except during initial radio-collaring or re-radio-collaring episodes) and therefore had few or no observations

Table 5.1. *Exposure frequencies and sample sizes of disease-causing organisms in ocular swab samples from both eyes of mountain lions (*Puma concolor*) sampled in the Black Hills, South Dakota, 2006–2009*

	% Positive ($n_{positive}/n_{total}$)			
Sex	Feline herpes virus	*Mycoplasma* spp.	*Chlamydophila felis*	Feline calicivirus
Female	1.8 (1/57)	1.8 (1/55)	0.0 (0/55)	5.5 (3/55)
Male	2.6 (1/38)	0.0 (0/30)	0.0 (0/30)	3.3 (1/30)
Both sexes	2.1 (2/95)	1.2 (1/85)	0.0 (0/85)	4.7 (4/85)

Source: Jansen 2011.

Table 5.2. *Percent of mountain lions with positive titers for disease-causing viruses in blood sera in the Black Hills from 2006–2009*

		% Positive						
Sex	Age	FIV	FLV	FHV	FCV	F/CPV	CDV	Plague
F	SA	0			11.1	55.6	0	
F	A	15.2			34.2	97.4	15.8	
F	All[a]	14.0	0.0	0.0	30.6	85.7	16.3	0.0
M	SA	0			26.7	73.3	0	
M	A	22.2			18.8	93.8	43.8	
M	All[a]	10.3	2.6	2.8	22.2	75.0	19.4	0.0
All	All[a]	12.5	1.0	1.2	27.1	81.2	17.6	0.0

Source: Jansen 2011.
Notes: Age: SA = subadult, 1–3 yrs. old; A = adult, >3 yrs. old; All = both ages combined.
Viruses: FIV = feline immunodeficiency virus, FLV = feline leukemia virus, FHV = feline herpes virus, FCV = feline calicivirus, F/CPV = feline/canine nonspecific parvo virus, CDV = canine distemper virus, Plague = *Yersinia pestis*.
[a]Includes small number of kittens (<1 yr. old).

indicating lions were ill or that could be used to document the progression of disease within individual lions.

Another factor that increased our awareness of the implications of disease in mountain lions was a suit brought against the state of South Dakota by the Mountain Lion Foundation of California. A witness for the plaintiff stated that mountain lions could harbor diseases that, in concert with harvest, would drive the population to extinction. At the time, we had no information about disease exposure and assumed that these lions were healthy based on external characteristics and nutritional condition information obtained during necropsies (see chapter 6). However, our experience with cloudy-eye syndrome increased our suspicion that diseases, some associated with domestic species, began to affect mountain lions in the Black Hills as the species expanded into areas occupied by humans (and vice versa), likely from lions killing and consuming domestic pets (i.e., cats and dogs). For example, exposure to feline immunodeficiency virus (FIV) (Biek et al. 2006) averaged 12% in mountain lions in the Black Hills (table 5.2). This disease also is associated with domestic cats and has been documented in lions throughout Montana (25% of 352 lions [Biek et al. 2006]) and in Washington (Evermann et al. 1997). It has been suggested that both mountain lions and African lions (*Panthera leo*) had been exposed to FIV longer than domestic cats had been, providing a long period of coevolution between pathogen and host. What was interesting was that this long period of evolution might explain the lack of pathogenic effects of the disease on large felids like mountain lions (Carpenter et al. 1996; Culver 2010). Therefore, despite the implications that this disease might contribute to the mortality of mountain lions, there likely are no effects of it that could contribute to population decline and extirpation of lions within the Black Hills region. In any case, the exposure rate was rather low compared to that in other mountain lion populations.

More telling was that 81% of the sampled mountain lions showed titers for feline/canine nonspecific parvo virus, and 18% showed exposure to canine distemper virus (table 5.2). In addition, a much stronger association between feline calicivirus and cloudy-eye syndrome was seen with the results of serum analyses: 27% of mountain lions in the Black Hills were exposed to the disease—as mentioned, the estimated proportion of mountains with this eye disease was between 5% and 10%. Although not confirmatory, such a high percentage of positive animals was unexpected at this time. These findings further implicate interactions with domestic cats and dogs as the source of at least some of the exposure to these diseases.

During our studies of mountain lions, we did not collect information on clinical signs of diseases such as feline leukemia; collection of clinical characteristics requires frequent interactions with animals as well as laboratory analyses to confirm diseased lymph nodes, low red blood cell count, and septicemia. However, feline leukemia was not documented in lions in California (Paul-Murphy et al. 1994) or in early work on Florida panthers (Cunningham et al. 2008). Exposure to feline leukemia was low (about

1% of tested samples). Therefore, although deaths of Florida panthers had been associated with the disease (Brown et al. 2008), we did not expect it to negatively affect the population of mountain lions inhabiting the Black Hills.

We also documented low exposure to feline herpes virus, a disease that can cause conjunctivitis. This disease was of interest, since mountain lions at the time were affected by cloudy-eye syndrome and 1–3% of lions did show titers for exposure to the disease. As with other potential pathogens, if cloudy-eye syndrome resulted from lion-specific interactions with infected domestic species, feline herpes virus, together with feline calicivirus, could explain the prevalence of cloudy eyes in the Black Hills mountain lion population. It also could explain the disappearance of the disease once the population began to decline because of harvest. Again, we were unable to verify the causative agent associated with this eye malady, and therefore the suggestion of a feline herpes virus effect is circumstantial at best. Furthermore, although 19% of 58 mountain lions tested for feline herpes virus in California showed titers indicating exposure (Paul-Murphy et al. 1994), eye characteristics indicating cloudiness were not reported.

The death of a National Park Service employee in Grand Canyon National Park from exposure to plague (*Yersinia pestis*) (Paul-Murphy et al. 1994), bobcat exposure to plague in North Dakota (D. M. Fecske, North Dakota Game and Fish Department, unpublished data), and the fact that the disease had been documented in black-tailed prairie dog (*Cynomys ludovicianus*) towns in southwestern South Dakota, prompted us to test for exposure to plague and proceed cautiously when conducting necropsies of mountain lion mortalities. Of the 68 mountain lions tested for exposure, none was positive. Those results were interesting, since the disease was documented within the region, and since lions immigrating to the Black Hills from west of the region, where the disease was more common, could have exposed Black Hills lions to the disease. However, because exposure to the disease is believed to increase with age (allowing more time for lions to come in contact with infected prey) and could be higher in female than in male mountain lions (Biek et al. 2006), our evaluation of this disease may have preceded its entry into the Black Hills system.

Another disease organism that has been associated with eye inflammation (fig. 5.3) is toxoplasma (e.g., *Toxoplasma gondii*), a parasitic protozoan (Dardé, Ajzenberg, and Smith 2011). In fact, that disease is closely associated with felids, which are the only known definitive hosts, and because it is common in domestic cats, it is of concern with regard to previously unexposed pregnant women. Because of this close association between the disease and cat species, and because we documented lions feeding on domestic cats, we expected that tests for exposure of lions to the disease would be positive. Of 97 mountain lions tested, 53% were positive for exposure (table 5.3). Other species also can carry this disease, and therefore, mountain lions could have become exposed via consumption of domestic pigs, for example. Further, there could be regional variation in mountain lions' susceptibility to the disease, based on the

FIGURE 5.3. Despite limited vision caused by cloudy-eye syndrome, affected mountain lions were able to survive the condition until their vision improved. *Photo by Dan Thompson.*

*Table 5.3. Exposure frequencies and sample sizes for disease-causing organisms in blood sera of mountain lions (*Puma concolor*) sampled in the Black Hills, South Dakota, 2006–2009*

Sex	Age	*Toxoplasma* IgM	*Toxoplasma* IgG	*Bartonella* IgG
F	All[a]	0.0	56.9	5.2
M	All[a]	0.0	46.2	10.3
All	All[a]	0.0	52.6	7.2

Source: Jansen 2011.
[a]Includes small number of kittens (<1 yr. old).

distribution of prey susceptible to the protozoan. Rotstein et al. (2000) stated that prevalence varied from 18% for Florida panthers to 28% for mountain lions from Texas that were released in Florida.

We also tested for *Bartonella* spp., otherwise known as "cat scratch fever" (Rotstein et al. 2000). In fact, exposure was twice as high in male lions as in female lions (table 5.3). Scratches obtained during territorial interactions of male lions can result in exposure, and depending on the number of wounds, death can result from the disease. One such subadult mountain lion that fought with a territory holder of approximate age of 4.5 years died from wounds to the head and front paws (see fig. 4.2 [top]). The territory holder had bitten off both ears, sliced through the skin about the eyes and forehead, and bitten the front forelegs multiple times (there were more than 20 lacerations to the right front foreleg of the younger lion). Although undocumented, a high probability of disease transmission was mentioned when the case was discussed with disease pathologists at SDSU.

We did not quantify internal (e.g., helminths) and external (e.g., ticks) parasites of mountain lions when conducting necropsies, even though the species is known to carry a number of these pathogens (Forrester, Conti, and Belden 1985). We were unaware of these external parasites partially because most of the carcasses we necropsied that had not been previously skinned were generally free of these parasites. Some mountain lions that were necropsied, especially older individuals, held numerous tapeworms, and it seemed as if the quantity increased or that internal parasites were more common in lions necropsied around the time we believed the population became saturated, as well as when consumption of deer by lions became more common. As mentioned, *Taenia omissa* was prevalent in 37% of mule deer from the southern Black Hills (Zimmerman 2004); this species of tapeworm has been documented in Florida panthers (Foster et al. 2006) and in mountain lions in Manitoba (Dare and Watkins 2012). Rausch, Maser, and Hoberg (1983) documented nine *Taenia* species from 39 mountain lions obtained from northeastern Oregon. Furthermore, *Taenia omissa* were documented in a mountain lion that died in Connecticut and was linked to the Black Hills (Hawley et al. 2016). Therefore, it is likely that the internal parasites we observed in mountain lion intestines also were *Taenia*.

Mange (*Notoedres cati*) has been documented in mountain lions (Uzal et al. 2007), and the prevalence of this disease can increase with population size; however, we did not document the typical alopecia (hair loss) and skin crusts normally associated with mange in any of the mountain lions necropsied, either when the population was believed to be low or after it had increased. Again, most carcasses that we necropsied had been skinned, and thus, some could have been infected with this disease, but our field observations of captured lions did not provide support for its existence.

Our experiences with diseases provided some fascinating information on the potential relationship between mountain lions and diseases that are mostly associated with domestic species. Our experiences with cloudy-eye syndrome suggest that infectious diseases that become problematic with high population size may become trivial once the population is reduced to a size that minimizes interactions between lions, humans, and their pets. Nevertheless, as we have documented with dispersing mountain lions (Thompson et al. 2009), domestic pets will be consumed by inexperienced younger lions in need of an easy meal.

Literature Cited

Bauman, P. J., J. A. Jenks, and D. E. Roddy. 1999. Evaluating techniques for monitoring elk movement across fence lines. Wildlife Society Bulletin 27:344–352.

Biek, R., T. K. Ruth, K. M. Murphy, C. R. Anderson Jr., M. Johnson, R. DeSimone, R. Gray, M. G. Hornocker, C. M. Gillin, and M. Poss. 2006. Factors associated with pathogen seroprevalence and infection in Rocky Mountain cougars. Journal of Wildlife Diseases 42:606–615.

Brown, M. A., M. W. Cunningham, A. L. Roca, J. L. Troyer, W. E. Johnson, and S. J. O'Brien. 2008. Genetic characterization of feline leukemia virus from Florida panthers. Emerging Infectious Diseases 14:252–259.

Carpenter, M. A., E. W. Brown, M. Culver, W. E. Johnson, J. Pecon-Slattery, D. Brousset, and S. J. O'Brien. 1996. Genetic and phylogenetic divergence of feline immunodeficiency virus in the puma (*Puma concolor*). Journal of Virology 70:6682–6693.

Culver, M. 2010. Lessons and insights from evolution, taxonomy, and conservation genetics. In Cougar: Ecology and conservation, ed. M. Hornocker and S. Negri, 27–40. Chicago: University of Chicago Press.

Cunningham, M. W., M. A. Brown, D. B. Shindle, S. P. Terrell, K. A. Hayes, B. C. Ferree, R. T. McBride, et al. 2008. Epizootiology and management of feline leukemia virus in the Florida puma. Journal of Wildlife Diseases 44:537–552.

Dardé, M. L., D. Ajzenberg, and J. Smith. 2011. Population structure and epidemiology of *Toxoplasma gondii*. In *Toxoplasma gondii*: The model apicomplexan: Perspectives and methods, ed. L. M. Weiss and K. Kim, 49–80. London: Academic Press / Elsevier.

Dare, O. K., and W. G. Watkins. 2012. First record of parasites from cougars (*Puma concolor*) in Manitoba, Canada. Canadian Field-Naturalist 126:324–327.

Evermann, J. F., W. J. Foreyt, B. Hall, and A. J. McKeirnan. 1997. Occurrence of puma lentivirus infection in cougars from Washington. Journal of Wildlife Diseases 33:316–320.

Forrester, D. J., J. A. Conti, and R. C. Belden. 1985. Parasites of the Florida panther (*Felis concolor coryi*). Proceedings of the Helminthological Society Washington 52:95–97.

Foster, G. W., M. W. Cunningham, J. M. Kinsella, G. McLaughlin, and D. J. Forrester. 2006. Gastrointestinal helminthes of free-ranging Florida panthers (*Puma concolor coryi*) and the efficacy of the current anthelmintic treatment protocol. Journal of Wildlife Diseases 42:402–406.

Hawley, J. E., P. W. Rego, A. P. Wydeven, M. K. Schwartz, T. C. Viner, R. Kays, K. L. Pilgrim, and J. A. Jenks. 2016. Long-distance dispersal of a subadult male cougar from South Dakota to Connecticut documented with DNA evidence. Journal of Mammalogy 96:1435–1440.

Jacques, C. N., J. A. Jenks, A. L. Jenny, and S. L. Griffin. 2003. Prevalence of chronic wasting disease and bovine tuberculosis in free-ranging deer and elk in South Dakota. Journal of Wildlife Diseases 39:29–34.

Jansen, B. D. 2011. Anthropogenic factors affecting mountain lions in the Black Hills, South Dakota. PhD diss., SDSU.

Jenks, J. A., D. M. Leslie Jr., and H. C. Gibbs. 1986. Anomalies of the skull of a white-tailed deer fawn from Maine. Journal of Wildlife Diseases 22:286–289.

Logan, K. A., and L. L. Sweanor. 2001. Desert puma: Evolutionary ecology of an enduring carnivore. Washington, DC: Island Press.

Maehr, D. S. 1997. Florida panther: Life and death of a vanishing carnivore. Washington, DC: Island Press.

Paul-Murphy, J., T. Work, D. Hunter, E. McFie, and D. Fjelline. 1994. Serologic survey and serum biochemical reference ranges for free-ranging mountain lion (*Felis concolor*) in California. Journal of Wildlife Diseases 30:305–315.

Rausch, R. L., C. Maser, and E. P. Hoberg. 1983. Gastrointestinal helminthes of the cougar, *Felis concolor* L., in northeastern Oregon. Journal of Wildlife Diseases 19:14–19.

Rotstein, D. S., S. K. Taylor, J. Bradley, and E. B. Breitschwerdt. 2000. Prevalence of *Bartonella henselae* antibody in Florida panthers. Journal of Wildlife Diseases 36:157–160.

Thompson, D. J., D. M. Fecske, J. A. Jenks, and A. R. Jarding. 2009. Food habits of repatriating cougars in the Dakotas: Prey obtained from prairie and agricultural habitats. American Midland Naturalist 161:69–75.

Uzal, F. A., R. S. Houston, S. P. D. Riley, R. Poppenga, J. Odani, and W. Boyce. 2007. Notoedric mange in two free-ranging mountain lions (*Puma concolor*). Journal of Wildlife Diseases 43:274–278.

Waid, D. D., and D. B. Pence. 1988. Helminths of mountain lions (*Felis concolor*) from southwestern Texas, with a redescription of *Cylicospirura subaequalis* (Molin, 1860) Vevers, 1922. Canadian Journal of Zoology 66:2110–2117.

Zimmerman, T. J. 2004. Effects of fire on the nutritional ecology of selected ungulates in the southern Black Hills. Master's thesis, SDSU.

CHAPTER 6

Nutritional Ecology of Mountain Lions

My experience with wildlife nutrition began at the University of Maine, where I worked under the direction of Dr. David M. Leslie Jr. and raised deer fawns to assess the effects of digestibility of poor-quality winter diets on their nutrition. Then, while working on my PhD, I evaluated how cattle affect the nutritional condition of free-ranging deer in Oklahoma and Arkansas (Jenks and Leslie 2003). At the time, the techniques I used for deer were rarely evaluated in carnivores, likely because of the low number of carcasses available for assessing nutritional condition. However, those techniques were beginning to show up in the literature, and relationships being documented for the more common deer species were similar to those for, for example, the wolf (*Canis lupus*) (LaJeunesse and Peterson 1993). It seemed, therefore, a logical extension to evaluate nutritional condition in mountain lions when carcasses began to increase in frequency in the Black Hills.

Mountain lions are true carnivores, in the sense that they consume only meat and associated carcass components (bones, cartilage, fat). Because of their monotypic proclivity for diets high in protein, their requirements for nutrients differ from those of pseudocarnivores, such as coyotes and bears, which seasonally consume high amounts of fruit or other non-protein-rich food sources (for the coyote, see MacCracken and Uresk 1984; for the black bear, see Mosnier, Ouellet, and Courtois 2008). Using standard procedures for evaluating diets, it is rare to document a lion that has consumed foods other than what are considered standard prey sources. For example, a young male that has become independent from its mother and has thus begun to disperse may depart from the usual diet. In this situation and on a few other occasions, we have documented mountain lions consuming some vegetation. However, we were unsure whether the vegetation (pine needles, a few blades of rough grasses) was consumed while grooming, or because the animal had not eaten for an extended period of time, or because the animal was bored. These subadult lions, however, were thin, with low fat levels. One such observation was of a subadult male that died on Interstate 90 where a herd of elk commonly crossed the highway. It is possible that the male was in pursuit of these elk and was killed by a vehicle when attempting to cross the road. The lion had consumed a few blades of grass but nothing more.

One nutrient need that differentiates cats from other predators is the need for the amino acid taurine (Hedberg, Dierenfeld, and Rogers 2007). This nutrient aids in the development and functioning of the eye and, thus, the development of sight in cats. If deficient in taurine, kittens can become blind and/or develop sight-related problems. Concentrations required for normal development and function range between 0.04% and 0.2% of the diet, and intake is generally adequate when the animal consumes whole prey, such as rodents (Hedberg, Dierenfeld, and Rogers 2007). When we first documented cloudy-eye syndrome in mountain lions, we believed that the population was saturated. If lions at some saturation level have difficulty capturing prey because of reduced availability or begin consuming what might be considered low-quality prey, then, we hypothesized, taurine availability could become compromised and could lead to subsequent eye or sight-related problems. However, we were unable to link this disease with taurine availability or determine the disease vector responsible for the condition. The relationship between nutrition and disease for large free-ranging feline carnivores has yet to be evaluated.

We collected data on nutritional characteristics of mountain lions in the Black Hills from late 1999 through mid-2013. We worked closely with the South Dakota Department of Game, Fish and Parks, which at times received calls regarding kittens that had been abandoned, either owing to harvest, removal of an adult female, or death of the mother caused by vehicles or natural causes. When kittens were extracted from the wild, they would be transported to our departmental captive facility at SDSU (fig. 6.1). While they were in that facility, we would periodically collect information on body weight and age; the South Dakota Department of Game, Fish and Parks worked to place the kittens in zoos accredited by the Association of Zoos and Aquariums. We generally held captive lions less than about seven months, a period that allowed for collection of data and scheduling of transport to zoos. Over the time of our work (2003–2012), we placed kittens in zoos located in various cities, including Philadelphia, Salt Lake City, Kansas City, Denver, and Phoenix. While raising these kittens, I discussed appropriate foods for young cats with zoo personnel, who strongly suggested that these animals receive adequate taurine from dietary fish or some other component.

Information collected on captive kittens was combined with data on wild kittens of known age (up to 1 year) to estimate the rate of gain in weight for young lions in the Black Hills (fig. 6.2). Based on the combined data, we found that body weights of young lions can increase between about 0.50 kg and 0.62 kg per week ($n = 29$ kittens) in the Black Hills. Captive data were less variable than data on wild lions that were

FIGURE 6.1. Mountain lion kittens raised in captivity at South Dakota State University. Body weights were used to generate growth equations for young mountain lions. *Photo by Gail Jenks.*

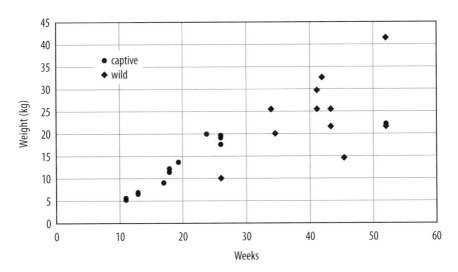

FIGURE 6.2. Body weights of mountain lion kittens raised in captivity (circles) and those of mortalities of young lions from the Black Hills (diamonds).

older than about 26 weeks. However, that would be expected, because as kittens aged, we fed complete diets for felids obtained from zoos, and also fresh chicken. These captive kittens gained 0.67 (±0.03) kg per week, significantly more than wild kittens, which gained 0.57 (±0.05) kg per week. Furthermore, using only kittens that maximized growth, the predicted weight gain ($r^2 = 0.99$) was 0.73 (±0.02) kg per week, which could represent the maximum weight gain for mountain lion kittens under ideal conditions. In fact, based on this model, the maximum weight of dispersing mountain lions would range between 35.3 kg (79.1 lbs.) and 39.67 kg (88.9 lbs.), which was comparable to the actual weights of 12 to 14 month old male lions leaving the Black Hills.

Male and female weights ($n = 239$ lions) plateaued around 1 to 3 years for females and 3 to 5 years for males (fig. 6.3). The weights of lions that were established were generally around 40 kg (88 lbs.) for females and 60 kg (130 lbs.) for males. For males, these weights were above those of dispersing animals (38 kg [84 lbs.]) and indicated that they had gained weight after establishing territories. The maximum weights of males surpassed 72 kg (158 lbs.), while females varied from about 40 kg after establishing a home range to about 50 kg (110 lbs.) when pregnant. Male growth rates were 2.6 times those of females, whereas female growth rates were about 22% more variable than those of males.

We did notice some senescence with a 13-year-old female that weighed 39 kg (fig. 6.3). However, a female aged at 9–10 years when it was collared later successfully raised a litter, indicating that, given sufficient prey, the species can continue to con-

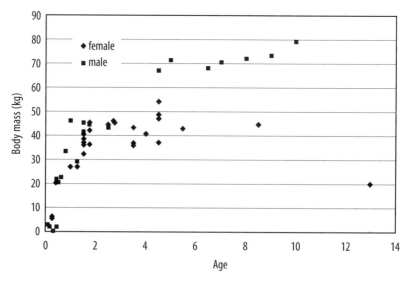

FIGURE 6.3. Body weights of male and female mountain lions (aged to 13 years) from the Black Hills.

tribute young to the population. Furthermore, one female lost her litter to infanticide during summer, only to become pregnant again and give birth to a litter in December, indicating that females could obtain sufficient food to regain fat reserves, become pregnant, and then give birth about 94 days later (a relatively short period of time for a large mammal).

Mountain lions are considered stalking predators. To maximize their chances of success when attempting to capture prey, they take advantage of thick vegetation as hiding cover, or they may attack from up-slope of prey to gain an advantage when a quick sprint is needed to successfully capture and kill prey (Fecske 2003). Lions have low endurance relative to chases, so prey that can run fast and long likely will evade capture. As with any predator, most attempts at capture fail, but as animals learn the habits of their prey, they can improve their chances of success. Thus, older mountain lions likely have higher success rates than subadults. Fat stores might be a good indicator of predator success because when less energy is required for stalking and capturing prey, the fat can be stored for later use.

When we first began necropsying mountain lions, we used a simple ranking system of low, medium, and high to assess fat reserves. Lions classed as having low fat had little to no fat around the kidneys and virtually no mesentery fat (abdominal fat associated with the exterior of the stomach and intestines). Carcasses classed as having medium fat had some kidney fat (the kidneys could be viewed with some exterior fat and there was some mesentery fat) (fig. 6.4). Those lions classed as having high fat exhibited extensive fat encasing the kidneys to the extent that they could be completely covered and difficult to locate, and mesentery fat was extensive.

In addition to ranks, we collected data on both Riney (1955) and total kidney fat indices (Finger, Brisbin, and Smith 1981), which have been used to evaluate the nutritional condition of white-tailed deer (Kie, White, and Drawe 1983; Jenks and Leslie 2003, 2011) as well as other large cervids. These indices are based on the amount of fat immediate to the kidney (Riney kidney fat index [RKFI]; see Riney 1955) plus fat associated with the kidneys but extending beyond the perpendicular ends of the organ (total kidney fat index [TKFI]; see Finger, Brisbin, and Smith 1981). These techniques involve weighing fat and kidneys; Riney and total kidney fat weights were divided by the weight of the kidneys (multiplied by 100). As expected, the fat indices were highly correlated ($r^2 = 0.61$) with each other (fig. 6.5) and were similar between males and females ($P = 0.47$); estimates of RKFI and TKFI averaged 54.7% and 127.4% for females and 70.9% and 148.5% for males, respectively.

Independent analyses also provided support that these values were stable relative to sex and age (table 6.1), although we did document a year effect, which supported use of the technique for evaluating population-level changes that were potentially related to change in population size. The change in fat levels through time also supported our hypothesis that population saturation (i.e., capacity of the Black Hills to support mountain lions) occurred around and after 2005 (table 6.1). These first estimates of fat

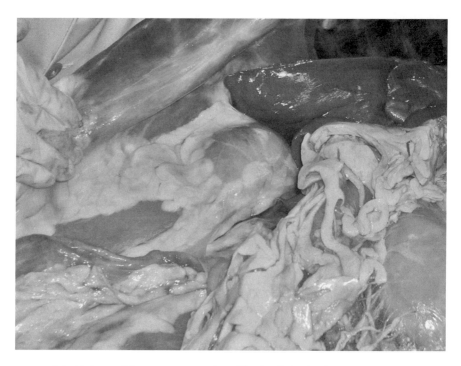

FIGURE 6.4. Abdominal fat in a mountain lion. This lion's fat level was visually ranked as moderate to high, because there was a vast amount of mesentery fat, but the kidneys were visible (not completely covered with fat). *Photo by D. Fecske.*

stores also allowed us to quantify our visual estimates of fat levels, because we had both ranked fat stores as low, medium, or high and had weighed fat and kidneys to estimate Riney and total kidney fat indices (table 6.2).

Our ranks for fat accumulation based on visual observations differed significantly from one another relative to total kidney fat; average TKFI percentages were 40.30% ($\pm$15.88%, $n = 11$) for low, 132.45% ($\pm$11.23%, $n = 22$) for medium, and 215.89% ($\pm$17.56%, $n = 9$) for high ($P < 0.001$) (table 6.2). Based on this analysis, we used values for TKFI when further analyzing fat levels of mountain lions relative to temporal trends.

Dispersing males and injured lions that no longer could capture and kill prey would generally have low fat levels; both RKFI and TKFI would be close to 0%. In contrast, pregnant females and prime-age adult males had high levels of fat (commonly surpassing 200% TKFI). Estimates for TKFI were variable across ages of mountain lions (fig. 6.6), with levels low for young lions; fat levels generally peaked at 1–3 years of age. Above age 1, TKFI percentages declined ($P = 0.02$) with age for males but not ($P = 0.36$) for females. Estimates of TKFI were more variable for lions older than

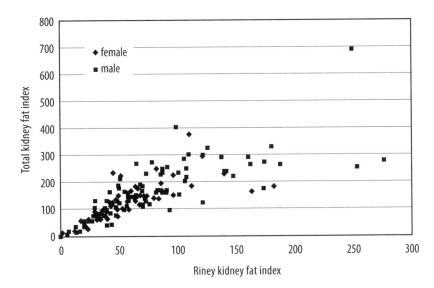

FIGURE 6.5. Relationship between the Riney kidney fat index and the total kidney fat index for mountain lion carcasses from the Black Hills. As expected, the two indices were correlated ($r^2 = 0.61$) with each other and similar ($P = 0.47$) between males and females.

Table 6.1.   Mean (and standard error) total kidney fat indices of adult and young mountain lions in the Black Hills

Period	Age	Mean	SE	N
1998–2003	<2.5 years	167.59	17.73	14
1998–2003	≥2.5 years	161.21	20.01	11
2004–2005	<2.5 years	118.54	13.83	23
2004–2005	≥2.5 years	114.18	19.15	12

Note: Total kidney fat is presented because Riney kidney fat indices did not differ ($P > 0.05$) among visual ranks for abdominal fat.
Age × Year: $P = 0.37$
Age: $P = 0.59$
Year: $P = 0.07$

Table 6.2.   Average total kidney fat indices (average and standard error of the mean) for mountain lions from the Black Hills

Code	Average TKFI (%)	SEM	N
L = Low	40.30	15.88	11
M = Moderate	132.45	11.23	22
H = High	215.89	17.56	9

Notes: Lion carcasses were ranked low, medium, and high based on condition of abdominal fat.
Total kidney fat is presented because values for Riney kidney fat did not differ ($P > 0.05$) among visual categories.
$F = 27.83$; $df = 2,39$; $P < 0.001$

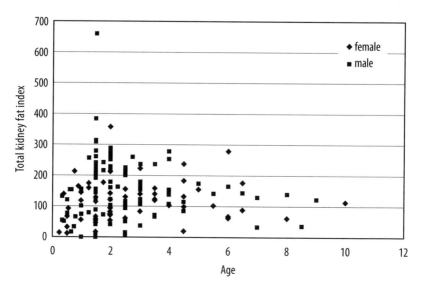

FIGURE 6.6. Relationship between total kidney fat index and age and sex of mountain lions from the Black Hills.

5 years, especially for males, which possibly reflects the costs of holding and maintaining territories. TKFI for females, although less variable than for males, surely reflects the costs of reproduction. For example, despite 13 years of age, one female had relatively high fat levels, suggesting that sufficient food was available for fat storage, even though the body weight of this individual had declined.

When looking across time, we did document some variation in average TKFI levels that occurred after 2005–2006, when the population was believed to be saturated (fig. 6.7). Means for males and females were closely associated with each other prior to this year and subsequently became disjunct, with females averaging generally lower fat reserves; this was the period linked to cloudy-eye disease in some lions. Interestingly, mean percentages for TKFI for males and females became closely associated again in 2013, after the population had been reduced from saturation levels by harvest. At that time, expectations were that prey availability would have increased per individual lion. Because fewer males would be moving and competing for territories, and sufficient prey was available to support these fat reserves, female reproductive effort might have become more stable as well, although the effects of kitten mortality through infanticide also could contribute to these values.

As can be seen, studying a species over a long term allows for an understanding of changes affecting individuals and populations temporally. We were able to follow this population from just after colonization, or from a low population size, to what we believe was population saturation, to a reduced population size owing to a manipulated

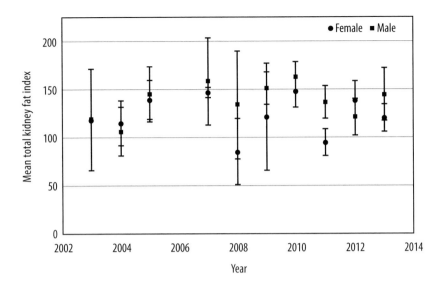

FIGURE 6.7. Temporal change in mean total kidney fat indices of male and female mountain lions from the Black Hills. Note the change in pattern after about 2005 that suggests that total kidney fat declined and was more variable, especially in females, until 2012. Years 2012–2013 indicate that fat levels became similar to 2005 levels relative to amount and variability between males and females.

reduction by harvest. During this time we also were able to document subtle changes in foraging habits that likely represented how prey use changed with predator population size. For example, we confirmed a strong association between mountain lions and porcupines early in our studies that indicated that lions had significantly reduced porcupine availability in the Black Hills. Alternatively, Sweitzer (1996) documented significantly higher predation on porcupines by mountain lions when the primary prey of lions, mule deer, declined in northwestern Nevada. Therefore, porcupines likely represent an important prey during initial colonization of areas and later when other prey become restricted.

One interesting finding related to capture of porcupines was the perceived method lions used to consume the prey. When necropsying carcasses after lions had consumed porcupines, we would find whole paws of the prey in the stomach, along with quills, leg bones, and muscle tissue. It seems that lions removed the paws with their long nails after orienting the prey on its back (this pattern also was observed for consumed domestic cats, mink, turkeys, and badgers [D. Thompson, Wyoming Game and Fish Department, pers. commun.]). From that position, and with the porcupine's paws absent, lions then consumed the porcupine from the abdomen, leaving most of the quills with the skin (fig. 6.8). Nevertheless, when a mountain lion consumed porcupines, quills were found in the lion's front legs and chest, muzzle, tongue, stomach, and intestines. About the time the lion population became saturated, we documented increased consumption of domestic pets (cats and dogs). Hence, lions had moved from what was considered an easy-to-kill wild prey item to an easy-to-kill domestic prey item (fig. 6.8; table 6.3).

Some evidence that might support how establishing mountain lions respond to prey availability was collected from animals that died while traversing the prairies of North and South Dakota (Thompson et al. 2009). We were able to necropsy 14 carcasses of animals that had died from harvest (legal or illegal), vehicles, removal because of a problem animal by South Dakota Department of Game, Fish and Parks, an electrocution of a lion that climbed a power-line pole, and a nontarget snaring (table 6.3). These lions ranged in age from about 1 to 4 years. Of the total, 87% showed evidence of interactions with and consumption of porcupines, but only 50% were consuming deer, and some (14%) had consumed domestic cats. These findings closely parallel our early work on lions located in the Black Hills and support this opportunistic pattern of foraging by lions when they travel through unfamiliar habitat. Dispersing lions have been linked to riparian zones while traveling across unfamiliar territory, so it should not be surprising that this opportunistic predator would take advantage of potential prey in these areas. Thus, we also documented consumption of beaver (*Castor canadensis*), mink (*Neovison vison*), and badger (*Taxidea taxus*) by these dispersing animals (table 6.3).

FIGURE 6.8. Remains of a porcupine killed and consumed by a mountain lion in the Black Hills. *Photo by D. M. Fecske.*

Table 6.3. Stomach and gastrointestinal (GI) tract contents of Dakota mountain lions, 2003–2007

ID	Age	Cause of Death	Stomach/GI Tract Contents
F1	3.0–4.0 yrs.	Harvest	Beaver, rodent, PPQ[a]
F2	3 yrs.	Vehicle	Trace amount of deer hair
F3	2.0–3.0 yrs.	Vehicle	Deer, PPQ[a]
F4	1.0–2.0 yrs.	Legal Kill	White-tailed jackrabbit, rodent, PPQ[a]
M1	1.5–2.5 yrs.	Harvest	Deer, vegetation, PPQ[a]
M2	2.0–2.5 yrs.	Legal Kill	Domestic housecat, PPQ[a]
M3	1.0–2.0 yrs.	Vehicle	Deer, porcupine (in stomach), vegetation, PPQ[a]
M4	1.5–2.5 yrs.	Legal Kill	Deer
M5	2 yrs.	Legal Kill	Badger, mink, deer, PPQ[a]
M6	2 yrs.	Harvest	Deer, porcupine (in stomach)
M7	2.5–4.0 yrs.	Illegal Kill	Vegetation, PPQ[a]
M8	2.0–3.0 yrs.	Removal	Domestic housecat, PPQ[a]
M9	3 yrs.	Electrocution	Empty, PPQ[a]
M10	3.0–4.0 yrs.	Incidental Snare	Empty, PPQ[a]

Source: Thompson et al. 2009.

[a]Mountain lion carcass contained prevalence of porcupine quills (PPQ) imbedded in the flesh at time of necropsy.

Table 6.4. Number of necropsied mountain lions that had evidence of porcupine consumption (%Porcupine), domestic animal consumption (%Domestic), and were classed as emaciated (%Emaciated), based on carcass condition (fat content and external characteristics), 1998–2005, in the Black Hills

Year	N	Age	%Porcupine	%Domestic	%Emaciated
1998	2	2	50	0	0
1999	6	2.2	100	0	0
2000	3	3	50	0	0
2001	3	1	34	0	0
2002	10	2.5	70	0	0
2003	10	2	55	33	0
2004	20	2	63	25	10
2005	25	2	35	20	4

At the time when the lion population was believed to be saturated (after about 2005), the South Dakota Department of Game, Fish and Parks received numerous reports of lions eating domestic cat and dog food that had been placed outside doors or on patio decks; and lions were observed near homes, presumably looking for food. Lions also were found in garages or outbuildings (potentially hunting mice) or barns (in close proximity to livestock). Mountain lions were documented killing llamas, sheep, and goats, and there were reports (although most were unverified) of attacks on cattle and horses. In fact, over 20% of necropsied lion carcasses showed evidence of consumption of domestic species (table 6.4). In response to these kills and reports, the

South Dakota Department of Game, Fish and Parks advised residents to keep pets indoors, to keep livestock in pens or barns, to keep domestic animal areas lighted, and to remove vegetation close to residences; these recommendations have also been presented to residents in other states to deter lions from encountering humans and domestic animals.

In response to the increased number of mountain lion reports and information collected during our first studies of mountain lions in the Black Hills, the South Dakota Department of Game, Fish and Parks had initiated a harvest, but because of the uncertainty both in the population information collected at that time and because some South Dakota residents were not in favor of a harvest, a low harvest limit was established. In addition, the harvest was classed as "experimental" in that the information collected during the first two years would be used to evaluate the effect of harvest on this population of mountain lions. Results from the initial harvests collected via radio-collared animals and population modeling indicated that the low harvest limit was not affecting the population to any real extent. We suspected that the population was responding to the low harvest just as any other species would do: increase survival as new subadult males established in the territories of harvested males and as female subadults established in the ranges of harvested females (fig. 6.9). We did not see other changes that would suggest that the population was reduced to any extent relative to the dynamics of individuals inhabiting the Black Hills.

However, as stated, there were changes in feeding habits. Mountain lions seemed to respond to the change in prey availability with the reduction in numbers of porcupines and to the change in behavior of Black Hills residents in response to the killing of their pets and domestic animals. At the individual level, we had seen lions that killed deer early in our studies, but they seemed to be focused on mule deer, not white-tailed deer, despite the high availability of white-tails and low availability of mule deer (table 6.5). We had already documented that coyotes were preying on white-tailed deer in the northern and central Black Hills regions (Griffin et al. 1994, 1999; DePerno et al. 2000). At the time of those studies, we did not document any killing of white-tailed deer by mountain lions, even though we had more than 40 deer radio-collared in the northern Black Hills and 70 radio-collared in the central Black Hills. In the southern Black Hills, because of the diversity of habitats that favored both white-tailed deer and mule deer, we captured and radio-collared both species (Griffin et al. 2004). We saw differential mortality on the two prey species, with most mortality on mule deer attributed to mountain lions, whereas most mortality on white-tailed deer was attributed to coyotes.

When considering that lions likely colonized the southern Black Hills initially and that these lions had dispersed to the Black Hills from the west, where the primary prey of most, if not all, lion populations at the time was mule deer, it seemed logical that these first inhabitants of the Black Hills would be skilled at killing mule deer and that porcupines were an easy prey no matter where they occurred. White-tailed

FIGURE 6.9. Lactating adult female mountain lion in the Black Hills. *Photo by Emily Mitchell.*

Table 6.5. Cause-specific mortality of radio-collared white-tailed deer and mule deer in the southern Black Hills

	White-tailed deer		Mule deer	
	N	%	N	%
Harvest	2	4	1	6
Vehicles	7	14	0	0
Illegal harvest	3	6	1	6
Mountain lion	2	4	9	56
Coyote	18	36	4	25

Source: Griffin, Jenks, and DePerno 2004.

deer, to the contrary, are a more wary species when compared to mule deer, and white-tailed deer occupied more open habitats, ponderosa pine forests, many of which had been thinned by the USDA Forest Service to promote tree growth and minimize disease (i.e., mountain pine beetle [*Dendroctonus ponderosae*] infestation). Thus, these forests were relatively open and would promote visual identification of lions attempting to approach white-tails. In contrast, mule deer inhabited mountain mahogany and juniper scrub habitats with dense ground cover. These habitats were located in the foothills, where mule deer used established trails along hillsides. Use of these trails likely promoted the stalking and attack of mule deer by mountain lions, because of the close proximity between predator and prey (lions could hide in the thick grasses adjacent to trails). Based on classic behavior, lions would pounce on the prey and attempt to strangle it, either at the throat or by grasping the nasals or nose of the deer. On a number of these carcasses, the nasals had been bitten and removed from the facial area either during the attack or after the prey had been killed. This behavior seemed consistent for other large prey; I recall a mountain goat carcass found between Crazy Horse Mountain and Mount Rushmore north of Custer that had similar characteristics to these preyed-upon deer. Once the prey was killed, the lion would drag the carcass downslope into the thick vegetation at or near the bottom of a draw. At times, we documented lions spending as long as four days at these sites, revisiting the kills and covering the cached kills until most of the carcass had been consumed.

As indicated in table 6.5, mountain lions did kill white-tailed deer early in the recolonization of the region (or when the lion population was at low density), but data collected at that time indicated that the kill rate of this species was likely low compared to other prey. It was hypothesized that this predator that was new to the Black Hills had to learn how to kill this new but ubiquitous prey species and that the learning curve might have been hastened by the lowered availability of "easy" prey (porcupines) and the more limited availability of prey (mule deer, distributed mostly along the perimeter of the Black Hills region) that mountain lions were skilled at killing. Furthermore, there was that opportunistic character of mountain lions, which would allow for unusual or rare prey to be taken just because it was in the wrong place at the wrong time (for the prey) and was encountered. That might have been the case for a gray jay (*Perisoreus canadensis*), an American mink (*Neovison vison*), and a yellow-bellied marmot (*Marmota flaviventris*), all of which were documented during necropsy of mountain lion carcasses as having been consumed. Because mountain lions spend time in riparian zones dominated by white spruce (*Picea glauca*) during the daytime in summer, small mammals such as southern red-backed voles (*Myodes gapperi*) also have been consumed.

One historical factor occurring in the early years after 2000 that could be linked with the transformation from easy-to-kill prey to predation on larger deer and elk was that chronic wasting disease (CWD) was documented in captive elk in 1997–1998 (Jacques et al. 2003). In 2001 a free-ranging white-tailed deer that was harvested in

Fall River County near Hot Springs, South Dakota, also tested positive for the disease. In 2002 Wind Cave National Park personnel documented CWD in a Rocky Mountain elk within the park boundary and became concerned about distribution of the disease within resident populations of deer as well as in the elk population. In November 2002 a suspected 5-year-old female elk in Wind Cave National Park was euthanized and tested; this animal also was positive for CWD (Wind Cave National Park press release, November 18, 2002). By 2005 Schuler (2006) estimated that the prevalence rate for the disease in mule deer averaged about 9% and could have been as high as 15%. Since both mountain lions and CWD colonized the southern Black Hills initially, and since diseased prey likely represented easy kills, they could have contributed to the lions' transition from preying upon porcupine and domestic species to preying upon deer- and elk-sized prey.

Behavioral characteristics of deer and other cervids that are associated with CWD include the loss of fear of humans (and likely of predators), nervousness, and hyper-excitability (Schuler 2006). Furthermore, there is some indication that diseased individuals are shunned by healthy members of their species. If so, then diseased individuals would be located farther away from other individuals, and because of reduced vigilance and nervousness, they would be vulnerable to attack by lions. In fact, statistical modeling conducted by Schuler (2006) indicated that deer with CWD had a higher probability of being killed by a predator (in this case coyotes and mountain lions) than deer that were negative for CWD, even though coyotes were more likely than lions to leave tissues for disease sampling (table 6.6).

Overall, as the population approached saturation, we did see more deer in stomachs of necropsied lion carcasses. In addition, the South Dakota Department of Game, Fish and Parks provided funding specifically to evaluate mountain lion feeding patterns. We therefore modified our methods to collect data specifically on kill sites and to

Table 6.6. Cause-specific mortalities of deer at Wind Cave National Park, South Dakota, February 2003 to December 2005, and number of samples testable and nontestable for CWD

Cause	Testable (CWD positive)	Nontestable
Coyote[a]	3 (1)	2
Mountain lion	1 (0)	4
Vehicle collision	4 (1)	0
Hunting outside park	4 (0)	1
Lethal removal in park	5 (3)	0
Unknown	3 (2)	3

Source: Schuler 2006.
[a]Coyotes were more likely (60%) than mountain lions (20%) to leave tissues for disease testing.

document prey carcasses at those sites to determine what was being killed and how often lions were making kills (Smith 2014). For this work we concentrated our efforts on capturing adult males and females that had established home ranges (usually animals over 2.5 years of age), because of the cost of the technology (we used global positioning system [GPS] radio collars in place of traditional VHF radio collars) and because we were interested only in what lions that were resident to the Black Hills were consuming (fig. 6.10). Fortunately, the GPS technology was available, but unfortunately, the first collars we purchased had high failure rates, in part from damage that occurred when lions were capturing prey, fighting with other lions, mating, and engaging in other activities associated with general lion behavior. Later GPS collars (manufactured by Advance Telemetry Systems) were more robust relative to surviving normal mountain lion activities. We continued to experience failures, but not near the rate of the failures with previous collars. We also noted that most of the work conducted on feeding habits of mountain lions using GPS collars involved visiting locations that were called cluster sites (where clusters of GPS points [≥2 points within 200 m] indicated a potential kill at the site) after a considerable amount of time had elapsed between the kill and the site visit and where the programmed collars collected

FIGURE 6.10. A mountain lion fitted with a global positioning system radio collar (manufactured by North Star LLC) in the Black Hills. *Courtesy of B. Jansen.*

Table 6.7. Seasonal frequency of prey items (%) found for mountain lions in the Black Hills

Prey type	Adult F Summer (n = 548)	Adult F Winter (n = 462)	Adult M Summer (n = 101)	Adult M Winter (n = 108)	Subadult Summer (n = 158)	Subadult Winter (n = 129)
BH sheep	0.18	0.87	0.99	0.93	1.27	0.00
Bison	0.00	0.00	0.99	0.93	0.00	0.00
Coyote	0.55	1.74	0.00	1.85	0.63	2.33
Deer	86.68	85.90	72.28	64.81	77.22	88.37
Dom.	1.10	1.08	1.98	8.33	2.53	0.78
Elk	4.20	3.04	18.81	19.44	1.90	2.33
Lion	0.18	0.00	1.98	0.93	0.00	0.00
Marmot	0.91	0.00	0.00	0.93	2.53	0.00
Misc.	0.00	1.08	0.00	0.00	0.00	0.78
Porcup.	0.18	0.22	0.00	0.00	3.80	0.00
Turkey	3.28	2.60	0.00	0.93	6.33	4.65
Unkn.	0.73	0.87	2.97	0.00	0.63	0.78
Other	2.01	2.82	0.00	0.93	3.16	0.00
Scav.	7.30	23.43	18.81	38.89	10.13	27.13

Source: Smith 2014.

just a few locations per day. To deal with these potential issues, we decided to increase the number of locations per day collected by the GPS collars, because initial work indicated that we might be expending considerable time visiting sites where lions were just loafing rather than sites where actual "kills" or "feeding" on kills were in progress (Knopff et al. 2009; Smith 2014).

Most studies of mountain lion feeding habits in western states have found that up to 80% of the diet was composed of deer (Spalding and Lesoski 1971; Toweill and Meslow 1977; Ackerman, Lindzey, and Hemker 1984), and because these studies have been conducted in areas where mule deer are the common species (Logan and Sweaner 2001), most if not all of the diet is that species of deer (Robinette, Gashwiler, and Morris 1959). This was the case in the North Dakota Badlands (Wilckens et al. 2016). The Black Hills, on the other hand, house deer populations dominated by white-tailed deer; the ratio of the two species is about 4:1. Also, when there are two species of deer occupying the same area, it can be difficult to determine which species was actually consumed at kill or feeding sites, depending on the carcass condition during data collection. Therefore, when Smith (2014) compiled the data in table 6.7, they were presented as "deer," but the category was mostly white-tailed deer. These data were collected during a period when the lion population had been reduced by harvest and the number of nonharvest carcasses had declined from previous years, when the population was believed to be saturated. In comparison to previous periods, diet diversity had increased, but certain dietary items, namely porcupine, had decreased, and killing of this species was mostly associated with subadult lions instead of all classes (e.g., adult males and females, females with kittens) of lions (although the sample size was much higher than during our earlier work).

Diet diversity (in this case, diet richness or the number of different species consumed) ranged from about 6 to 9 species across females, males, and subadult lions (Smith 2014). Another notable finding, although not unique to the Black Hills (see Robinette, Gashwiler, and Morris 1959; Lindzey 1987), was that lions were killing other predators; both lions and coyotes were found at kill sites. Only a few of these predators were found, compared to the primary prey of the region, deer, which would suggest that these kills were rare and likely related to territorial disputes (with other lions) or competition for food (coyotes killed when attempting to feed at lion kill sites). Females and subadult lions killed more small deer-sized prey than did adult males. Adult males were killing larger deer and elk-sized prey. These results make sense based on the difference in body size between adult males and other sex and age classes of lions, since the larger the prey item is, the higher the risk of injury to the predator.

Historical folklore suggested that lions only eat prey that they kill, but that belief had been dispelled by a number of studies that found mountain lions scavenging on carcasses, including our recent work in North Dakota (Wilckens et al. 2016). What was unique to the Black Hills, however, was the amount of scavenging by lions. As mentioned, we successfully captured some mountain lions by baiting them into sites where we would subsequently place snares around carcasses of deer and other species. We knew lions would scavenge and took advantage of this behavior. Other lion populations have been documented scavenging in the neighborhood of about 7% of the time (which we documented for lions that had recently recolonized the North Dakota Badlands [Wilckens et al. 2016]). However, scavenged carcasses were found at lion cluster sites up to 39% of the time in the Black Hills. Adult males scavenged most in winter and less in summer, whereas adult females and subadult lions scavenged less than males did but generally more than has been found for other lion populations. Scavenge sites ranged from carcass dump sites established by the South Dakota Department of Transportation to vehicle kills on roads, to carcasses left at hunting camps after removal of some meat and, if a male, the antlers. To be classified as a scavenged carcass, there had to be some evidence that the lion did not kill the animal (e.g., evidence of saw marks on bones or missing antler plates [Smith 2014; Wilckens et al. 2016]). Another interesting observation was that there were enough carcasses in some areas of the Black Hills that a lions could survive and even successfully reproduce by consuming scavenged carcasses alone. For example, one female lion with a broken leg survived for six months while feeding kittens by consuming carcasses. After her leg had healed, the female began making kills again (J. B. Smith, SDSU, pers. commun.).

In 2016 approximately 7,200 elk inhabited the Black Hills outside of Wind Cave National Park and Custer State Park, as estimated by T. Kirchenmann of the South Dakota Department of Game, Fish and Parks (T. Kirchenmann, pers. comm.). Compared to white-tailed deer (approximately 40,000 animals [Cudmore 2017]), which are fairly uniformly distributed across the region (although not mule deer versus white-tailed deer), elk and other large prey tend toward more clumped distributions. Therefore, you might expect some lions to become better able to capture these clumped prey, because the lions' territories and home ranges overlap those of these less distributed and thus more rare prey species. Because the lions are more likely to encounter these species, they would gain experience in attempting capture, especially if they began by capturing young of the species (e.g., elk calves). If so, then certain lions would potentially focus their prey capture skills on deer, others on elk, and others on bighorn sheep, just to name the top prey available to lions in the Black Hills. In fact, some have argued that this proclivity for specific prey could mean that lions would eradicate some herds of bighorn sheep that had been released as part of restoration projects (Festa-Bianchet et al. 2006). At times we observed kills of elk calves and lambs of bighorn sheep, which would be about the same size as deer and thus would be more susceptible to predation once lions had become adept at capturing deer. However, adult female and male elk are considerably larger than deer, so the risk of injury during capture attempts would be much higher. Furthermore, other studies of lion populations have documented deaths of lions from injuries obtained during such capture attempts (Lindzey 1987). Therefore, the question came to mind, Why risk injury attacking adult elk when you could more easily capture deer and when, since deer are more prevalent, you have more experience capturing deer than those large, rare elk?

Chronic wasting disease could have come back into play with these larger kills, because lions living in the southern portions of the region would have had diseased elk and deer available to them. I received reports that hunters in Custer State Park (immediately north of Wind Cave National Park) had observed multiple dead elk carcasses in an area where there had been sightings of mountain lions. Hunters at the time were implicating lions in these deaths; they thought the predators were engaging in "surplus" killing. But why risk injury killing more than one elk, when it would take a considerable amount of time to consume the first one killed? We hypothesized that these sightings were more likely related to disease deaths, in this case CWD, and that the lions would be expected to select sick or diseased prey at higher proportions than healthy prey (Smith 2014), even though evidence for this phenomenon has been mixed (Barber-Meyer, White, and Mech 2007; Krumm et al. 2009).

We had the opportunity to evaluate the selection of diseased elk by two adult mountain lions (one male, one female) that used portions of Wind Cave National Park (Smith 2014). From January 2011 to June 2012, we documented prey remains at 31 GPS cluster sites for the two lions in the park. For these two lions, prey items consisted of deer, turkey, coyote, one scavenged bison (*Bison bison*), and 20 elk carcasses. There were a total of 14 mountain lion–killed elk (7 by the male, 7 by the female) with tissue samples available to test for CWD (testing was conducted by Wind Cave National Park personnel). Six elk kills were not tested for the disease, but two of them were elk calves, which would not be expected to test positive. The average age of these kills was 9.8 years, and age was similar for those carcasses that were positive and those that were negative for the disease. Of the 14 elk killed by lions, 9 (64%) were positive for CWD. Although the sample size was low, these results supported selection of diseased over healthy elk. Based on these findings, we concluded that the elk carcasses observed by hunters were more likely related to disease deaths than to surplus killing by lions.

That is not to say that multiple kill sites by mountain lions in close proximity to one another do not occur. In these instances, however, there may be multiple lions feeding on the same kill or kills. For example, six mountain lions in the Black Hills were observed feeding at the same time on a vehicle-killed deer (fig. 6.11). Of the lions at the site, there was one adult female (radio-collared) and her two kittens, a kitten that was not a littermate (a suspected orphan), one adult male (based on size alone), and one unknown animal. One animal was barely visible (lying behind the adult female on the right [fig. 6.11]) but in front of the tree trunk (B. Jansen, SDSU, pers. commun.).

The long-term data set that we developed from necropsying mountain lion carcasses provided a pattern of prey use for the species as the population increased in size and then subsequently decreased owing to the increase in harvest (fig. 6.12). The pattern likely characterizes how a large predator might first establish within an area, learn

FIGURE 6.11. Mountain lions consuming prey at night in the Black Hills. *Photo courtesy of Brian Jansen.*

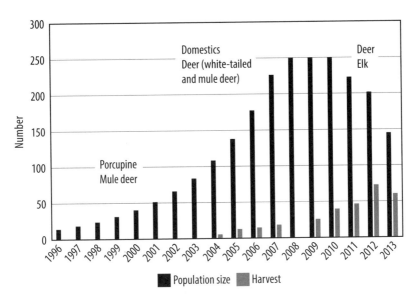

FIGURE 6.12. Population size and harvest of mountain lions in the Black Hills. Food habits changed based on ease of capture and knowledge of new prey, progressing from porcupine and mule deer in the 1990s to opportunistic captures of high-risk species including elk after 2010.

the behavior and distribution of new and possibly "naive" prey (Berger, Swenson, and Persson 2001), and become adept at capturing these prey as the population is managed to an equilibrium where sufficient prey are available to support the needs of those predators. Predators start with familiar prey and those that are easy to capture, learn how to capture prey that represent a relatively low risk of injury or death, and finally opportunistically capture more risky prey species, using behavioral cues linked to probability of success.

Literature Cited

Ackerman, B. B., F. G. Lindzey, and T. P. Hemker. 1984. Cougar food habits in southern Utah. Journal of Wildlife Management 48:147–155.

Barber-Meyer, S. M., P. J. White. and L. D. Mech. 2007. Survey of selected pathogens and blood parameters of northern Yellowstone elk: Wolf sanitation effect implications. American Midland Naturalist 158:369–381.

Berger, J., J. E. Swenson, and I. Persson. 2001. Recolonizing carnivores and naïve prey: Conservation lessons from Pleistocene extinctions. Science 291:1036–1039.

Cudmore, K. W. 2017. An evaluation of deer and pronghorn surveys in South Dakota. Master's thesis, SDSU.

DePerno, C. S., J. A. Jenks, S. L. Griffin, and L. A. Rice. 2000. Female survival rates in a declining white-tailed deer population. Wildlife Society Bulletin 28:1030–1037.

Fecske, D. M. 2003. Distribution and abundance of American martens and cougars in the Black Hills of South Dakota and Wyoming. PhD diss., SDSU.

Festa-Bianchet, M., T. Coulson, J. Gaillard, J. T. Hogg, and F. Pelletier. 2006. Stochastic predation events and population persistence in bighorn sheep. Proceedings of the Royal Society B 273:1537–1543.

Finger, S. E., I. L. Brisbin Jr., and M. H. Smith. 1981. Kidney fat as a predictor of body condition in white-tailed deer. Journal of Wildlife Management 45:964–968.

Griffin, S. L., J. A. Jenks, and C. S. DePerno. 2004. Seasonal movements and home ranges of white-tailed deer and mule deer in the southern Black Hills, South Dakota, 1998–2003. Completion Report No. 7583, Federal Aide to Wildlife Restoration, Job W-75-R-34. South Dakota Department of Game, Fish and Parks, Pierre, SD.

Griffin, S. L., J. F. Kennedy, L. A. Rice, and J. A. Jenks. 1994. Movements and habitat use of white-tailed deer in the northern Black Hills, South Dakota, 1990–1992. Completion Report, Federal Aide to Wildlife Restoration, Job W-75-R-33. South Dakota Department of Game, Fish and Parks, Pierre, SD.

Griffin, S. L., L. A. Rice, C. S. DePerno, and J. A. Jenks. 1999. Seasonal movements and home ranges of white-tailed deer in the central Black Hills, South Dakota and Wyoming, 1993–1997. Completion Report No. 99-03, Federal Aide to Wildlife Restoration, Job W-75-R-34. South Dakota Department of Game, Fish and Parks, Pierre, SD.

Hedberg, G. E., E. S. Dierenfeld, and Q. R. Rogers. 2007. Taurine and zoo felids: Considerations of dietary and biological tissue concentrations. Zoo Biology 26:517–531.

Jacques, C. N., J. A. Jenks, A. L. Jenny, and S. L. Griffin. 2003. Prevalence of chronic wasting disease and bovine tuberculosis in free-ranging deer and elk in South Dakota. Journal of Wildlife Diseases 39:29–34.

Jenks, J. A., and D. M. Leslie Jr. 2003. Effect of domestic cattle on the condition of female white-tailed deer (Odocoileus virginianus) in southern pine-bluestem forests, USA. Acta Theriologica 48:131–144.

Jenks, J. A., and D. M. Leslie Jr. 2011. Interactions with other large herbivores. In Biology and Management of White-Tailed Deer, ed. D. Hewitt, 287–310. Boca Ratan, FL: CRC Press International, Taylor and Francis Group.

Kie, J. G., M. White, and D. L. Drawe. 1983. Condition parameters of white-tailed deer in Texas. Journal of Wildlife Management 47:583–594.

Knopff, K. H., A. A. Knopff, M. B. Warren, and M. S. Boyce. 2009. Evaluating global positioning system techniques for estimating cougar predation parameters. Journal of Wildlife Management 73:586–597.

Krumm, C. E., M. M. Conner, N. T. Hobbs, D. O. Hunter, and M. W. Miller. 2009. Mountain lions prey selectively on prion-infected mule deer. Biological Letters 6:209–211.

LaJeunesse, T. A., and R. O. Peterson. 1993. Marrow and kidney fat as condition indices in gray wolves. Wildlife Society Bulletin 21:87–90.

Lindzey, F. 1987. Mountain lion. In Wild furbearer management and conservation in North America, ed. M. Novak, J. A. Baker, M. E. Obbard, and B. Malloch, 657–668. Toronto, Ontario: Ministry of Natural Resources.

Logan, K. A., and L. L. Sweanor. 2001. Desert Puma: Evolutionary ecology and conservation of an enduring carnivore. Washington, DC: Island Press.

MacCracken, J. G., and D. W. Uresk. 1984. Coyote foods in the Black Hills, South Dakota. Journal of Wildlife Management 48:1420–1423.

Mosnier, A., J. Ouellet, and R. Courtois. 2008. Black bear adaptation to low productivity in the boreal forest. Ecoscience 15:485–497.

Riney, T. 1955. Evaluating condition of free-ranging red deer (*Cervus elaphus*), with special reference to New Zealand. New Zealand Journal of Science and Technology 36:429–463.

Robinette, W. L., J. S. Gashwiler, and O. W. Morris. 1959. Food habits of the cougar in Utah and Nevada. Journal of Wildlife Management 23:261–273.

Schuler, K. L. 2006. Monitoring for chronic wasting disease in deer at Wind Cave National Park: Investigating an emerging epidemic. PhD diss., SDSU.

Smith, J. B. 2014. Determining impacts of mountain lions on bighorn sheep and other prey sources in the Black Hills. PhD diss., SDSU.

Spalding, D. J., and J. Lesoski. 1971. Winter food of the cougar in south-central British Columbia. Journal of Wildlife Management 35:378–381.

Sweitzer, R. A. 1996. Predation of starvation: Consequences of foraging decisions by porcupines (*Erethizon dorsatum*). Journal of Mammalogy 77:1068–1077.

Thompson, D. J., D. M. Fecske, J. A. Jenks, and A. R. Jarding. 2009. Food habits of repatriating cougars in the Dakotas: Prey obtained from prairie and agricultural habitats. American Midland Naturalist 161:69–75.

Toweill, D. E., and C. Meslow. 1977. Food habits of cougars in Oregon. Journal of Wildlife Management 41:576–578.

Wilckens, D. T., J. B. Smith, S. A. Tucker, D. J. Thompson, and J. A. Jenks. 2016. Mountain lion (*Puma concolor*) feeding behavior in the recently recolonized Little Missouri Badlands, North Dakota. Journal of Mammalogy 97:373–385.

CHAPTER 7

Genetics of Mountain Lions

As with any initial project working on a new species in a new area, when we began working on mountain lions in the Black Hills, we had to base all of our hypotheses on what was known about lions in other systems. Even armed with this information, we knew we were dealing with a species believed to have recently recolonized the Black Hills. Therefore, our immediate questions included those aimed at recolonizing populations. For example, we asked, Where did the first lions in the Black Hills come from? Were lions always in the Black Hills, but at such a low population size that they remained secretive and were rarely seen? If the population was low, would we expect it to suffer from genetic issues such as low heterozygosity or limited variability among alleles, either of which could make Black Hills lions somewhat unusual genetically and potentially vulnerable to abnormalities that have afflicted the Florida panther? Factors such as genetic drift and natural selection also could negatively affect a newly re-established population.

Our first captures of lions provided some indications that healthy individuals, both males and females, made up the Black Hills population. We documented individuals that did not display the usual genetic characteristics relegated to homozygous lions, such as undescended testes and crooked tails (Maehr 1997). Nevertheless, if the population had been extinct or recently recolonized, it could exhibit a "founder effect," which is seen when only a few individuals colonize or occur in a region and those individuals are responsible for all or most of the breeding, thus limiting genetic diversity in the population as it increases in size. If diversity was limited, then the population could suffer from susceptibility to disease or might decline at a faster rate than expected in response to some catastrophe that affected the population.

Some of our original questions were answered when we provided samples to the University of Wyoming, where a study of lion genetics was being conducted (Anderson, Lindzey, and McDonald 2004). That study supported our early observations of captured lions, that genetic diversity was sufficient for the maintenance of a healthy population. Nevertheless, our focus on the health, both genetic and physical (see chapters 5 and 6), of this new mountain lion population was increased by the establishment of a harvest season and the subsequent injunction against it filed by the Mountain Lion Foundations of California and of the Black Hills. Witnesses for the plaintiffs stated that the harvest would cause extinction of the species in the Black Hills and that the low population size was susceptible to diseases owing to low genetic diversity. That injunction, however, was rejected by the court in Pierre, South Dakota, which allowed the first limited harvest season on mountain lions to begin in October 2005.

However, we continued to wonder about the genetic diversity of the population, whether the harvest would affect the genetics of lions in the Black Hills, and whether the limited immigration of lions to the Black Hills documented by Anderson, Lindzey, and McDonald (2004), that is, one male per generation, was sufficient to maintain the genetic diversity of this population into the future. We therefore began collecting samples for genetic analyses from all captured lions and from those that were either found dead or reported to the South Dakota Department of Game, Fish and Parks and collected for future necropsy (Thompson 2009; fig. 7.1). Because this work was conducted immediately before and after initiation of the limited harvest, and the population at that time was believed to be increasing and approaching saturation, we believed that such an evaluation would provide an accurate indication of the species's ability to withstand the removal of individuals through either natural mortality or harvest.

Evaluating Genetic Diversity

From 2003 to 2006, we collected samples (muscle tissue and blood) from 134 mountain lions in the Black Hills. The previous analysis of genetic diversity (Anderson, Lindzey, and McDonald 2004) that included samples from lions in the Black Hills had used 8 samples at 9 microsatellite loci. Because we were interested in determining heterozygosity as an index not only of population health but also of the parentage of individuals, Dorothy (Fecske) Wells, at the time the furbearer biologist for North Dakota Game and Fish Department, was able to work with Michael Schwartz at the USDA Forest Service Genetics Laboratory, Missoula, Montana, to add to that total and thus allow the use of 20 microsatellite loci for our analyses (table 7.1). During this time, we also were able to add 18 mountain lions from North Dakota to our samples of lions from the Black Hills, which had increased substantially over the years, and not only determine genetic diversity of the North Dakota lions but also assess the genetic structure (compare the lion genetics of populations adjacent to those of the Black Hills to determine whether they were genetically distinct). We also conducted population assignment tests (individual lions are assigned to populations based on their genetics) among these semi-isolated populations. At the time, North Dakota

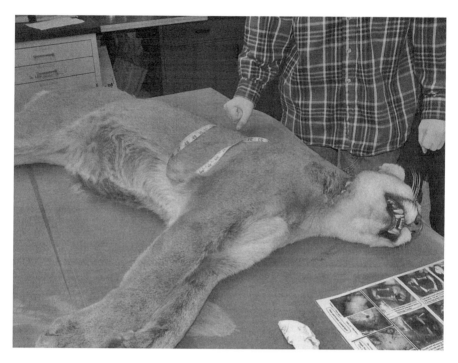

FIGURE 7.1. Tissue and blood samples were collected from mountain lions throughout our studies of the species in the Black Hills (2001–2013). *South Dakota Department of Game, Fish and Parks.*

also had initiated a harvest of mountain lions in the Badlands region of the state, where the population was believed to have recolonized the region as a result of the dispersal of lions from the Black Hills (Thompson 2009; Thompson and Jenks 2010). We were able to compare the genetics of these Dakota lion populations to the lion database from Wyoming (Anderson, Lindzey, and McDonald 2004), which, again, was originally based on 8 microsatellite loci (specific laboratory methodology can be found in Thompson [2009] and Juarez [2014]).

Our first look at lion genetics indicated that lions in the Black Hills averaged 4.3 alleles per locus and 86 alleles, levels associated with genetically diverse lion populations. First indications were that mountain lions in the Black Hills had more alleles than those from North Dakota, but this difference was likely due only to the large sample size of lions from the Black Hills compared to that for North Dakota. In fact, when these data were scaled based on sample size, effective alleles (table 7.1) were much more similar between the two populations, and for some loci, lions in North Dakota were more diverse. This result was our first indication that lions from outside the Black Hills were at least partially involved with colonizing the North Dakota

Table 7.1. Comparison of genetic variability by locus for samples taken from South Dakota (SD) and North Dakota (ND) mountain lions

Locus	N SD	N ND	Alleles SD	Alleles ND	Effective Alleles[a] SD	Effective Alleles[a] ND	H_O SD	H_O ND	H_E SD	H_E ND	F_{IS} SD	F_{IS} ND
Fca43	134	18	5	2	1.92	1.60	0.50	0.39	0.48	0.38	−0.05	−0.04
Fca57	133	18	5	4	1.99	2.93	0.45	0.72	0.50	0.66	0.09	−0.10
Fca77	134	18	2	2	1.01	1.06	0.01	0.06	0.01	0.05	0.00	−0.03
Fca90	134	18	5	5	2.38	2.78	0.60	0.67	0.58	0.64	−0.03	−0.04
Fca96	134	18	5	3	2.74	2.76	0.63	0.72	0.64	0.64	0.00	−0.13
Fca132	134	18	5	4	3.02	2.46	0.76	0.61	0.67	0.59	−0.14	−0.03
Fca559	132	18	8	6	3.42	1.81	0.70	0.28	0.71	0.45	0.02	0.38
Fca176	130	18	4	6	2.96	2.37	0.72	0.72	0.66	0.58	−0.09	−0.25
Fca35	133	18	2	2	1.92	1.86	0.44	0.50	0.48	0.46	0.09	−0.08
Lc109	134	18	4	2	2.91	1.53	0.71	0.33	0.66	0.35	−0.08	0.04
Fca391	133	18	4	3	2.34	2.11	0.53	0.44	0.57	0.53	0.07	0.16
Fca08	134	18	2	3	1.74	2.18	0.45	0.72	0.43	0.54	−0.05	−0.33
Fca30	134	18	4	2	1.29	1.25	0.22	0.22	0.22	0.20	0.04	−0.13
Fca82	121	18	6	3	2.98	2.99	0.61	0.78	0.66	0.67	0.08	−0.17
Fca149	134	18	3	3	1.41	1.48	0.27	0.39	0.29	0.32	0.08	−0.21
PcoA208	131	18	3	3	2.59	2.96	0.63	0.72	0.61	0.66	−0.03	−0.09
PcoB10	132	18	7	4	3.23	3.56	0.70	0.72	0.69	0.72	−0.01	0.00
PcoC112	133	18	4	2	2.57	2.00	0.60	0.53	0.61	0.50	0.01	−0.06
PcoB210	133	18	5	4	3.58	2.17	0.74	0.39	0.72	0.54	−0.03	0.28
PcoC108	133	18	3	3	2.86	2.62	0.67	0.61	0.65	0.62	−0.03	0.01

Source: Thompson 2009.
Notes: We assessed genetic diversity of mountain lions at 20 microsatellites (see Locus) for 134 samples from the Black Hills and 18 samples from North Dakota. Results presented here include alleles, effective alleles, observed heterozygosity (H_O), expected heterozygosity (H_E), and inbreeding coefficient (F_{IS}).
[a]Effective alleles represent the number of alleles by locus scaled by sample size for mountain lions in South Dakota and North Dakota.

Badlands. Other indices (observed and expected heterozygosity) also supported this finding.

Nevertheless, mountain lions in the Black Hills had an average expected heterozygosity (H_E) of 0.542 and an average observed heterozygosity (H_O) of 0.547, both of which were slightly higher than for lions in the North Dakota Badlands (table 7.2). So, despite the genetics indicating that lions from outside the Black Hills had contributed to the genetic diversity of North Dakota lions, the diversity of the North Dakota lions was a bit lower, possibly owing to the lower number of individuals in that population and/or just the lower sample size available at that time for North Dakota lions.

Both lion populations, however, had comparable or higher levels of genetic diversity when compared to other lion populations as well as to other carnivore species (Thompson 2009; table 7.3). Although some of these other studies suffered from analyses that encompassed a lower number of microsatellites, which could account for

Table 7.2. *Mean and standard error of genetic variability metrics for Dakota mountain lions*

	South Dakota		North Dakota	
	Mean	SEM	Mean	SEM
Alleles/locus	4.3	0.356	3.3	0.282
Allele with freq ≥5%	3.3	0.252	2.6	0.169
Effective alleles/locus	2.442	0.162	2.223	0.148
No. of alleles	1.3	0.291	0.3	0.147
Alleles exclusive to				
H_E	0.542	0.041	0.504	0.039
H_O	0.547	0.044	0.526	0.046

Source: Thompson 2009.

Table 7.3. *Comparison of observed heterozygosity (H_O) levels of mountain lions in the Black Hills with other mountain lion populations and other carnivore species*

Species	Region	H_O	Source
M. lion	Utah	0.47	Sinclair et al. 2001; 9 loci
M. lion	Western US	0.42–0.52	Culver et al. 2000; 10 loci
M. lion	California	0.36–0.42	Riley et al. 2014; 54 loci
Cheetah[a]	South Africa	0.39	Menotti-Raymond and O'Brien 1995
African lion[a]	Africa	0.66	Menotti-Raymond and O'Brien 1995
Leopard[a]	Africa	0.77	Spong, Johansson, and Björklund 2000
Brown bear[a]	Kodiak Islands	0.30	Paetkau et al. 1998
Gray wolf[a]	US	0.54	Roy et al. 1994

Source: Adapted from Thompson 2009.
[a]Cheetah (*Acinonyx jubatis*), African lion (*Panthera leo*), leopard (*Panthera pardus*), brown bear (*Ursus arctos*), and gray wolf (*Canis lupus*).

the lower levels of diversity, the indication was that, relative to other populations, the Dakotas (both the Black Hills and the North Dakota Badlands) had sufficient diversity to at least maintain genetically healthy populations in the short term, provided there was no cataclysmic factor that substantially reduced these populations quickly. It also could be noted that other carnivore populations had higher levels of genetic diversity than mountain lion populations in the Dakotas, and therefore there was a possibility that genetic diversity levels could increase above those presently documented (table 7.3).

Our findings on the genetics of mountain lions supported the original work by Anderson, Lindzey, and McDonald (2004) and provided analytical evidence that paralleled what our visual observations of captured and collected lions had indicated: that Black Hills lions were genetically diverse and that there did not seem to be any risk of depleting that diversity through the limited harvest that was initiated in 2005.

The other interesting finding obtained from this analysis was that the Black Hills population as well as the North Dakota Badlands population had unique alleles

(SD: $n = 26$; ND: $n = 6$), indicating that lions from other regions of North America contributed to the recolonization of both populations. That finding agreed with our expectations for mountain lions in the Black Hills, because we knew that individuals likely dispersed to the region from Wyoming, but the finding did not agree with our initial hypothesis for the lions that colonized the Badlands region of North Dakota. In that case, we had assumed that lions just continued to move north from the Black Hills, because the timing seemed to fit well with colonization first of the Black Hills and then of the Badlands.

This result also differed from statements made by some biologists involved with the Eastern Cougar Network, who had concluded that all mountain lions documented east of the Black Hills had come from the Black Hills region and thus that any reduction in Black Hills lions, through harvest or other factors, would reduce the natural recolonization of more eastern states. Nevertheless, our findings of unique alleles in both populations indicated that there was considerable movement of lions in the west (including lions in Montana, Wyoming, and even Colorado) that were dispersing to the east. Thus, multiple lion populations were involved with the recolonization of the Dakotas, and given the long dispersal distances that could be attained by lions, these alternative populations could represent the home areas of some of those first individuals that had headed even farther east to states such as Minnesota, Wisconsin, Iowa, Illinois, and Missouri.

Our analysis of both Dakota lion populations also provided information supporting the belief that the species had reestablished from just a few individuals that had "founded" these populations (Thompson 2009; Juarez 2014). Nevertheless, these mountain lion populations showed no deleterious effects from those few founding individuals. Considering that residents and recreationalists from both regions had begun seeing and reporting observations of mountain lions in the 1990s through the very early years of the 2000s and that mountain lions were believed to be at low densities at that time, this finding made sense. However, the "new" alleles and the high genetic diversity also implied that individuals from various states and populations had been moving into the region(s) during that time (most likely males, but potentially some females) and that those new individuals had contributed to the breeding populations for both regions of the Dakotas (fig. 7.2).

The use of a larger number of microsatellite loci than had been used in previous work also allowed us to use assignment tests that provide an indication of the origin of the individual animals. When these tests were conducted, samples from 2 of the 18 lions that had been collected by North Dakota Game and Fish, either from harvested lions or from other carcasses within the state boundaries, were actually assigned to the Black Hills population, meaning they were more similar, genetically, to Black Hills lions than to other lions in the North Dakota Badlands. However, these lions had died east of the Badlands region and thus were dispersing lions that for some reason had bypassed the Badlands region of western North Dakota. This finding was somewhat

FIGURE 7.2. Samples for genetic analyses also were collected from mountain lions captured in the North Dakota Badlands *Photo by Dave Wilckens.*

expected, based on the information we had collected from radio-collared lions that had first dispersed from the Black Hills to the Slim Buttes region of Harding County, South Dakota, which abuts North Dakota. One of those dispersers traversed diagonally across North Dakota, only to stop north and east of Grand Forks along the Red River (Thompson and Jenks 2010). The lion remained in that area for a short period and finally crossed the river and moved to the Minnesota-Manitoba border in the Roseau Wildlife Management Area (we were able to follow the path of this individual with the help of the Minnesota Department of Natural Resources). At the time I alerted the Manitoba Ministry of Natural Resources, hoping that its employees would attempt to find and subsequently follow the animal if it moved farther to the north. Unfortunately, that did not happen. However, genetic analyses further confirmed this pattern of movement north and east out of the Black Hills (Juarez et al. 2016).

We expected, based on the original analysis conducted by Anderson, Lindzey, and McDonald (2004), that lions in the Black Hills were more closely related to those in Wyoming ($F_{ST} = 0.024$ [index of population structure]) than to those in North Dakota ($F_{ST} = 0.043$), and our more complete analysis of mountain lion genetics also indicated this relationship. Overall, these results provided support for a movement conduit for

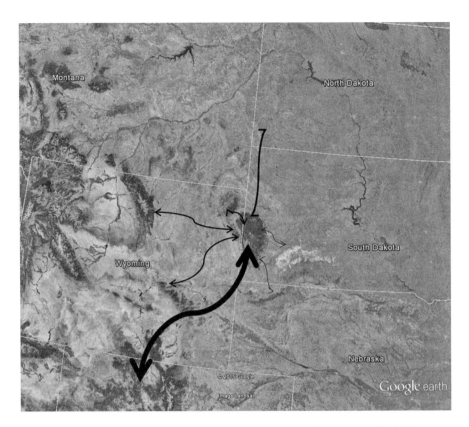

FIGURE 7.3. Conduit of movement of mountain lions into and out of the Black Hills, based on past and recent studies of genetics. This conduit is not absolute, since mountain lions from other populations to the west and potentially to the north contributed genetic material to both Black Hills and Badlands populations. Nevertheless, the conduit illustrates the degree of movement that occurs across the distribution of the species as well as the negligible effect of political boundaries on these patterns.

lions that began in the state of Wyoming along the Colorado border and continued through eastern Wyoming to the southern Black Hills (fig. 7.3).

As mentioned, while occupying the Black Hills, some dispersal-age lions would traverse the border of the Black Hills. Some headed along the western edge of the region and then left the Black Hills to head north (both northwest [based on radio-collared lions] and northeast [based on radio-collared lions and genetic analyses]). The genetic results seem to almost establish a boundary for these movements, but the unique alleles in lions inhabiting the North Dakota Badlands and other movements of lions out of the Black Hills would indicate that the conduit did not have absolute western or even eastern boundaries, and as a consequence, lions from other populations could either enter the system and establish at any point or in any population, or could move within the conduit to the north and east or the north and west.

We also estimated effective population size (N_e) of lions in the Black Hills at 27.9 animals (22.65–38.97; 95% CI) (Thompson 2009); effective population size is an indication of the number of individuals in the population that are involved with reproduction and thus gene transfer. In species that, like mountain lions, exhibit polygynous mating systems, effective population size can be low, because a few adult males monopolize females for relatively long periods of time. Our estimate was lower than expected, but since N_e can be 10% to 20% of the total population, which would indicate that between 138 and 279 mountain lions inhabited the Black Hills, the estimate of effective population size was in line with our estimates of total population size of mountain lions, generated from mark-recapture and population modeling.

In 2010, the potential for the use of genetics to improve understanding of this relatively new mountain lion population came up for discussion. We had been radio-collaring mountain lions for 10 years, and the South Dakota Department of Game, Fish and Parks was interested in using genetics with mark-recapture to ascertain whether this new technique would provide an estimate of population size comparable to the estimates obtained using the more standard marks, in this case radio-collared individuals. In addition, the department, through the Game and Fish Commission, had increased the harvest limit on the population, and there was evidence that the increased harvest had resulted in a reduction in population size. If there was some reduced population level that preserved genetic diversity, then the lower population size might be more manageable, while also preserving the genetic health of the population. Given these questions, we began this evaluation of the technique around 2012 (Juarez 2014).

To complete these analyses, we maintained our relationship with the USDA Forest Service Genetics Laboratory in Missoula, Montana, because we wanted to eliminate any potential for laboratory-specific effects (M. Schwartz, USDA Forest Service, Rocky Mountain Research Station, pers. commun.) that would result in inconsistencies between results from our first evaluation (Thompson 2009) and those from our new evaluation (Juarez et al. 2016). We reanalyzed those first results to ensure quality control of early and late samples using the original 20 microsatellite loci. We also were able to increase the number of samples obtained from adjacent populations, to improve our initial evaluation of population structure (how closely related these populations were, genetically, to one another), and generated a new estimate of effective population size (N_B) (individuals genetically contributing to the population) (Waples and Teel 1990; Waples 2005).

Our first question focused on the maintenance of genetic diversity of the lion population. For this revised analysis, we were able to include 675 mountain lions sampled from the Black Hills, 113 lions sampled from North Dakota, and 62 lions sampled from eastern Wyoming. For the Black Hills sample, we separated it into three periods, preharvest ($n = 288$), moderate harvest ($n = 289$), and heavy harvest ($n = 98$), which allowed an assessment of the effect of reduced population size on genetic diversity. During this time period (2001 to 2012), the harvest limit increased from 13 to 71 mountain lions per season. In contrast to the original analysis, we used a cohort analysis (Juarez et al. 2016), which provided estimates of genetic diversity per year from 2001 to 2012 (table 7.4). This improved cohort analysis indicated that observed heterozygosity ($P = 0.473$), expected heterozygosity ($P = 0.886$), allelic richness ($P = 0.764$), and effective alleles ($P = 0.745$) were statistically similar during the three periods when the population was believed to change in size (from low, to high, and then lower owing to harvest). Furthermore, the estimates of the effective population size of breeders (N_B) supported the predicted pattern of population change (table 7.4; fig. 7.4).

These results were interesting relative to understanding how many mountain lions were necessary to maintain this population in the long term (a viable population is

Table 7.4. *Estimates (at 20 loci) of observed (H_O) and expected (H_E) heterozygosity, allelic richness (Ar), and number of effective alleles (A_e) of Black Hills mountain lions by cohort year under different harvest regimes*

Harvest regime	Cohort year	N		H_O	H_E	Ar	A_e
Preharvest	2001–2002	56	Mean	0.56	0.55	4.07	2.49
(2001–2006)			SE	0.04	0.04	0.35	0.17
	2002–2003	76	Mean	0.57	0.56	4.02	2.57
			SE	0.04	0.04	0.34	0.18
	2003–2004	96	Mean	0.56	0.56	4.01	2.56
			SE	0.04	0.04	0.32	0.17
	2004–2005	117	Mean	0.57	0.55	3.95	2.51
			SE	0.04	0.04	0.31	0.16
	2005–2006	132	Mean	0.56	0.55	3.93	2.53
			SE	0.04	0.04	0.33	0.17
	2006–2007	149	Mean	0.57	0.56	4.01	2.60
			SE	0.05	0.04	0.31	0.18
Moderate harvest	2007–2008	146	Mean	0.56	0.56	3.94	2.57
(2007–2010)			SE	0.05	0.04	0.31	0.18
	2008–2009	142	Mean	0.54	0.55	3.86	2.52
			SE	0.04	0.04	0.32	0.18
	2009–2010	138	Mean	0.55	0.55	3.97	2.51
			SE	0.04	0.04	0.30	0.17
	2010–2011	110	Mean	0.55	0.55	3.98	2.51
			SE	0.04	0.04	0.30	0.17
Heavy harvest	2011–2012	79	Mean	0.55	0.56	3.90	2.57
(2011–2013)			SE	0.04	0.04	0.31	0.18
	2012–2013	52	Mean	0.54	0.55	3.99	2.50
			SE	0.04	0.04	0.33	0.17

Source: Juarez et al. 2016.

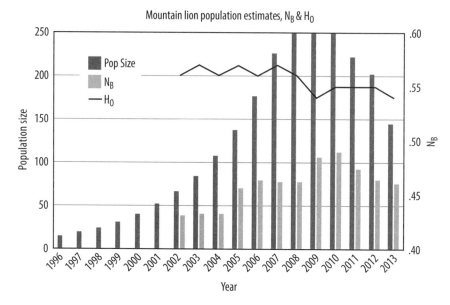

FIGURE 7.4. Population estimates (of the South Dakota portion of the Black Hills) of mountain lions derived from population simulation based on quantitative and qualitative data (Jenks, 2013 unpublished data), observed heterozygosity, and effective number of breeders (N_B) over time. *Juarez et al. 2016.*

considered to last 50 years or longer) (Allendorf and Luikart 2007). However, they did not provide a threshold lower level below which we would expect to see decreased genetic diversity and the consequences of increased homozygosity, reduced survival of offspring, and the more observable characteristics such as undescended testes and crooked tails. Nevertheless, our results did support the conclusion that genetic diversity could be maintained over multiple population sizes. For example, observed heterozygosity was constant despite significant changes in population size and numbers of effective breeders (N_B) (table 7.4). There were three years (2002, 2005, and 2013) in which the proportion of effective breeders was above 50% but estimated population size ranged from 66 to 138 to 145. This might suggest that the population could be conservatively reduced up to about 50%, and if the number of effective breeders was maintained above 50%, then genetic diversity would be preserved for at least the short term. However, such preserved genetic diversity assumes that adequate immigration would occur to continue the incorporation of new breeders into the system. Furthermore, the genetic preservation assumes that other variables (random mortality with loss of allelic diversity) would not hamper the population.

Our improved genetic analysis using additional samples from surrounding states and lion populations also indicated that South Dakota and eastern Wyoming lion populations were more closely related than South Dakota lions and those inhabiting the North Dakota Badlands. This result was expected, based on our original analysis (Thompson 2009) and because of the close proximity of mountain lions occupying the western portion of the Black Hills and Bear Lodge Mountains northwest of the Black Hills. Furthermore, the similarity in genetic variability between South Dakota and eastern Wyoming lions supported the "conduit" model (fig. 7.3), in that these adjacent populations separated by believed unsuitable (i.e., prairie) habitat were acting like a large panmictic population, as was originally suggested by Anderson, Lindzey, and McDonald (2004). There were, however, unique alleles in Wyoming ($n = 6$), the Black Hills ($n = 2$), and North Dakota ($n = 3$) lions, a finding that also supported the conduit of movement from Wyoming into the Black Hills as well as a uniqueness among these three populations (fig. 7.3). Our analysis confirmed the movement of six individuals from the Black Hills to North Dakota, but because of the additional samples, it noted movement to the south and west as well; two lions moved from North Dakota into the Black Hills.

Estimating Population Size

By 2011 we had radio-collared more than 300 mountain lions and had answered a number of questions about this population. At the time, the South Dakota Department of Game, Fish and Parks was interested in evaluating the use of genetics within a mark-recapture analysis to estimate population size. If such a technique would work, then the department could focus on obtaining tissue samples with biopsy darts rather than immobilizing lions, radio-collaring them, and monitoring their survival and movements weekly as well as annually. To address this question, we continued to radio-collar lions but merged the DNA analyses conducted to address genetic diversity questions with the radio-collared data we used to estimate population size (table 7.5). In these analyses we used only lions that were at least 2 years of age, to focus on animals that had established home ranges and territories in the Black Hills region. Thus we did not include subadult lions, which had a high likelihood (at times greater than 90% [Thompson and Jenks 2010; Jansen 2011]) of leaving the Black Hills once they began to disperse, thereby complicating our evaluation of genetics for the purpose of estimating population size.

We had a second issue to address in these analyses: would lions that were darted to obtain tissue well in advance of harvest still be alive during the harvest period? Others who had conducted this type of analysis (Beausoleil, Warheit, and Martorello 2005) had obtained tissue samples over a short period just prior to harvest, which would eliminate or reduce the potential for lion mortality and dispersal movements that would affect estimates of population size (both outcomes would elevate estimates of population size, because the marked lions would be thought to be available for harvest). Luckily, we had radio-collared lions to estimate survival over the period of

Table 7.5.   Radio and DNA-marked mountain lions (≥2 years of age) availability and harvest for 2012 and 2013

	2012				2013			
		DNA				DNA		
	Radio	NMPH	NMHH	NMTH	Radio	NMPH	NMHH	NMTH
Mountain lions available	33	32	32	31	40	28	27	26
Marked mountain lions harvested	11	7	7	7	9	8	8	8
Total harvested	63	63	63	63	50	50	50	50

Source: Juarez 2014.
Note: Three DNA-marked availability estimates were derived: natural mortality occurred only prior to the harvest season (NMPH), a midpoint in which natural mortality occurred during half of the harvest season (NMHH), and natural mortality occurred throughout the harvest season (NMTH).

tissue collection, and those estimates could be used to adjust the availability of DNA-marked lions in our analyses. We adjusted our estimates assuming that natural mortality occurred prior to harvest, during half of the harvest season, and throughout the harvest season, to provide an indication of the size of the effect (Juarez et al. 2016; table 7.5). These adjustments reduced the availability of DNA-marked lions from 32 to 31 in 2012 and from 28 to 26 in 2013 (table 7.5).

Estimates of the availability of marked lions (both radio-collared and DNA-marked) were then used to estimate population size (table 7.6). The estimates obtained from radio-collared mountain lions were 177 and 188 for 2012 and 2013, respectively, and were not significantly different from each other, although the pattern suggested the potential for a slight population increase. Alternatively, the estimates from DNA-marked lions averaged 259 and 157 for 2012 and 2013, respectively. These estimates of population size derived from DNA-marked mountain lions, similar to those based on radio-collared lions, were not significantly different from one another; however, they indicated that the population could have declined by about 100 individuals (table 7.6). This pattern of potential population decline, although again not statistically significant, was in line with our estimates of the number of breeders (N_B; table 7.5), which supported population decline, and it also provided support for the population model we used to estimate population size. In sum, the use of genetics to estimate population size seemed promising, even though we had many assumptions that would need to be addressed or eliminated (such as focusing significant effort on biopsy-darting mountain lions within a short period just prior to harvest, to maximize

Table 7.6. *Population estimates for mountain lions aged ≥2 years of age (i.e., independent of females) in 2012 and 2013*

	2012				2013			
		DNA				DNA		
	Radio	NMPH	NMHH	NMTH	Radio	NMPH	NMHH	NMTH
N	177	266	259	253	188	162	158	151
SE	36	72	70	68	44	39	37	35
Lower 95% CI	108	124	121	119	102	86	84	81
Upper 95% CI	249	408	397	387	273	238	231	220

Source: Juarez 2014.
Note: Estimates were derived using 2-sample Lincoln-Peterson with Chapman modification, from radio-marked and DNA-marked mountain lions. Three DNA-marked estimates were derived: where natural mortality occurred only prior to the harvest season (NMPH), a midpoint in which natural mortality occurred during half the harvest season (NMHH), and natural mortality occurred throughout the harvest season (NMTH).

sample size while minimizing the potential for movement out of the harvest zone or death from natural mortality prior to or during harvest).

Our ability to use the mark-recapture technique for estimating the population size of mountain lions in the Black Hills was hampered by the behavior of the species, trouble with maintaining enough marked animals during harvest seasons, and extensive movements of individuals, in addition to other statistical issues (i.e., variation in capture probabilities [Russell et al. 2012]). We had marked some mountain lions, especially juveniles but also some adults, that dispersed from the Black Hills region, as well as some that died of natural causes during the harvest season. Also, it was difficult to generate a reasonable sample of marked adult males owing to the skewed sex ratio of the species in the Black Hills (the ratio of males to females was estimated at 0.42:1); thus, estimates were generated for the population, even though sex-specific harvest rates likely differed. Relative to the secretive nature of mountain lion behavior, and despite comments we received suggesting that hunters were able to get to and harvest any lion at any time throughout the Black Hills, we wondered whether multiple captures to periodically replace radio collars, as well as other interactions with humans, might have enhanced the ability of marked lions to evade harvest, at least by some, maybe older, individuals. If so, the results of population analyses using these animals would have affected the estimates of population size on an annual basis and thus would have led to much variation; the results might have potentially inflated population estimates, because a lower number of marked animals would be removed from the population.

In contrast to the use of marked (i.e., radio-collared) mountain lions to estimate population size, the use of genetics provided supportive evidence that the mountain

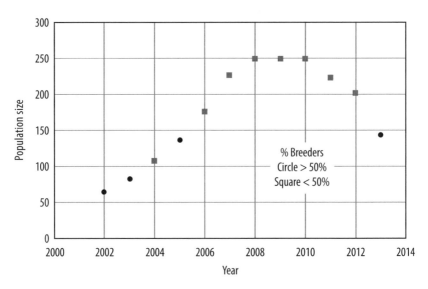

FIGURE 7.5. Change in percentage of effective breeders (NB) with population size of mountain lions in the Black Hills. Effective breeders were above 50% during periods of low and moderate population size. Mountain lions during all three periods showed high allelic diversity and heterozygosity. *Juarez et al. 2016.*

lion population in the Black Hills was healthy and diverse but likely had declined as a result of harvest. The population was benefiting from the immigration of individuals from western, long-established populations and even gained potential genetic diversity from individuals that had entered the Black Hills from the North Dakota Badlands. As a system of mountain lion movement, with the Black Hills representing a source of new individuals via colonizers that would disperse from the region mainly to the north and west but also to the east and south, the genetic diversity supported population and system viability. We used these comparative data to support our indices of population size and pattern of population change obtained through our mark-recapture analyses. These analyses, when combined with our estimates of numbers of effective breeders, suggest that, provided that the proportion of effective breeding individuals is close to or above 50% of the population (fig. 7.5), sufficient reproduction in conjunction with immigration can maintain genetic integrity of the mountain lion population in the Black Hills.

Our results on the genetics of mountain lions in the Black Hills also could be of interest relative to the present situation affecting the Florida panther. The size of the Black Hills is comparable to that of the area encompassing the habitat of the panther in southern Florida. However, population isolation in southern Florida has affected the genetic diversity of the panther, whereas in the Black Hills, even semi-isolation has allowed sufficient movement of genes into the population through immigration to maintain genetic diversity. Thus, provided that the conduit of movement of mountain lions is maintained, the population in the Black Hills should be able to conserve genetic diversity. Nevertheless, the temporal pattern of heterozygosity (fig. 7.5) could become problematic if a further decline occurs in the future. It would be beneficial to know the threshold population sizes that maintain genetic diversity with and without dispersal.

Literature Cited

Allendorf, F. W., and G. Luikart. 2007. Conservation and the genetics of populations. Malden, MA: Blackwell.

Anderson, C. R., Jr., F. G. Lindzey, and D. B. McDonald. 2004. Genetic structure of cougar populations across the Wyoming Basin: Metapopulation or megapopulation. Journal of Mammalogy 85:1207–1214.

Beausoleil, R. A., K. I. Warheit, and D. A. Martorello. 2005. Using DNA to estimate cougar populations in Washington: A collaborative approach. In Proceedings of the eighth mountain lion workshop, ed. R. A. Beausoleil and D. A. Martorello, 81–82. Olympia: Washington Department of Fish and Wildlife.

Culver, M., W. E. Johnson, J. Pecon-Slattery, and S. J. O'Brien. 2000. Genomic ancestry of the American puma (*Puma concolor*). Journal of Heredity 9:186–197.

Jansen, B. D. 2011. Anthropogenic factors affecting mountain lions in the Black Hills, South Dakota. PhD diss., SDSU.

Juarez, R. L. 2014. Estimating population size and temporal genetic variation of cougars in the Black Hills, South Dakota. Master's thesis, SDSU.

Juarez, R. L., M. K. Schwartz, K. L. Pilgrim, D. J. Thompson, S. A. Tucker, and J. A. Jenks. 2016. Assessing temporal genetic variation in a cougar population: Influences of harvest and neighboring populations. Conservation Genetics 17:379–388.

Maehr, D. S. 1997. Florida panther: Life and death of a vanishing carnivore. Washington, DC: Island Press.

Menotti-Raymond, M. A., and S. J. O'Brien. 1995. Evolutionary conservation of ten microsatellite loci in four species of Felidae. Journal of Heredity 86:319–322.

Paetkau, D., L. P. Waits, P. L. Clarkson, L. Craighead, E. Vyse, R. Ward, and C. Strobeck. 1998. Variation in genetic diversity across the range of North American brown bears. Conservation Biology 12:418–429.

Riley, S. P. D., L. E. K. Serieys, J. P. Pollinger, J. A. Sikich, L. Dalbeck, R. K. Wayne, and H. B. Ernest. 2014. Individual behaviors dominate the dynamics of an urban mountain lion population isolated by roads. Current Biology 24:1989–1994.

Roy, M. S., E. Geffen, D. Smith, E. A. Ostrander, and R. K. Wayne. 1994. Patterns of differentiation and hybridization in North American wolflike canids, revealed by analysis of microsatellite loci. Molecular Biology and Evolution 11:553–570.

Russell, R. E., J. A. Royle, R. DeSimone, M. K. Schwartz, V. L. Edwards, K. P. Pilgrim, and K. S. McKelvey. 2012. Estimating abundance of mountain lions from unstructured spatial sampling. Journal of Wildlife Management 76:1551–1561.

Sinclair, E. A., E. L. Swenson, M. L. Wolfe, D. C. Choate, B. Bates, and K. A. Crandall. 2001. Gene flow estimates in Utah's cougars imply management beyond Utah. Animal Conservation 4:257–264.

Spong, G., M. Johansson, and M. Björklund. 2000. High genetic variation in leopards indicates large and long-term stable effective population size. Molecular Ecology 9:1773–1782.

Thompson, D. J. 2009. Population demographics of cougars in the Black Hills: Survival, dispersal, morphometry, genetic structure, and associated interactions with density dependence. PhD diss., SDSU.

Thompson, D. J., and J. A. Jenks. 2010. Dispersal movements of subadult cougars from the Black Hills: The notions of range expansion and recolonization. Ecosphere 1:1–11.

Waples, R. S. 2005. Genetic estimates of contemporary effective population size: To what time periods do the estimates apply? Molecular Ecology 14:3335–3352.

Waples, R. S., and D. J. Teel. 1990. Conservation genetics of pacific salmon: Temporal changes in allele frequency. Conservation Biology 4:144–156.

CHAPTER 8

Perceptions of Mountain Lions

I have had many experiences interacting with wildlife professionals and recreational-ists interested in my research projects through the years. Before I began to study mountain lions, most of my projects were focused on deer nutrition, behavior, and population dynamics. People interested in those projects and findings included hunters, wildlife professionals, and employees of the USDA Black Hills National For-est, the latter likely because most of my projects were focused in the Black Hills. There was a change in who was interested as soon as I started to work on mountain lions. I began receiving many invitations to speak at sportsman shows and to make presentations to local wildlife groups such as the Brookings Chapter of the Wildlife Federation, university functions, and wild game banquets. At those venues I would present recent information we had been collecting on mountain lions, discuss some of the politics surrounding the species in South Dakota and beyond, explain how some misinformation had been corrected (e.g., it was learned that lions will consume carcasses), and teach people how to be safe when living and recreating in mountain lion country. After completing my presentations, I always allowed time for ques-tions and would stay around for attendees who wanted to share stories with me re-garding their personal experiences with the species. Many had seen mountain lions or thought they had seen one (fig. 8.1). Others were concerned about the safety of their children or pets, whether they lived within the Black Hills region or whether they frequently visited western South Dakota.

On a number of occasions, a contact who lived in the Black Hills volunteered to assist with a few of our early captures of mountain lions. He would help carry equip-ment and help with holding lions for data collection, including the weighing of animals. He always was positive and happy to assist with our work. However, his attitude toward lions became less favorable about the time the lion population became saturated

FIGURE 8.1. Adult as well as young mountain lions were observed by residents and visitors to the Black Hills region, especially when the population became saturated. *Photo by Dan Thompson.*

in the Black Hills. During that time, we had a phone conversation in which he stated that in the past he had frequently observed deer in the area surrounding his home, including his yard, but now he had been seeing few or no deer. And he had seen a lion pass through his yard, where deer had previously congregated. His friends, other hunters, had told him of a hunter who reported seeing a lion in Custer State Park in an area with multiple unconsumed elk carcasses and concluded that the one lion had killed all of those elk (that would have been surplus killing) (Kruuk 1972; DelGuidice 1998). The most important point he was trying to make to me was that lions were killing all the deer and elk in the Black Hills, and he and his friends were now unable to harvest these species because there were fewer harvest tags available.

I listened intently to his concerns and then began responding to his points. I explained how the elk population in the Black Hills had been reduced: because of damage claims by local landowners, additional elk harvest tags had been issued and additional animals taken by hunters; also, in Custer State Park, some elk likely had chronic wasting disease (CWD), which might explain the multiple elk carcasses he mentioned. I told him that some deer in the Black Hills also had CWD, especially in the southern region, which could affect the availability of this species as well. Despite multiple attempts to help him understand the observations he had mentioned, each time I stopped talking, he stated, "You don't understand, lions are killing all the deer and elk and I didn't get a tag this year." By the end of our conversation, it became obvious to me that the more simplistic conclusion would be the accepted cause, no matter the complexity of the situation, and when a large predator, in this case a mountain lion, was involved, the conclusion was going to be: predators, in this case mountain lions, kill prey, and more predators kill more prey. Another understanding was that there is a threshold of acceptance of large predators when they colonize or recolonize new regions.

The Early Years

In 2002 Larry Gigliotti, Dorothy Fecske (my first student working with mountain lions), and I collaborated on a public opinion survey regarding perceptions of South Dakotans regarding lions (Gigliotti, Fecske, and Jenks 2002). Prior to the survey, it was assumed that most South Dakotans pretty much hated and feared the new predators and wanted them eliminated from the state. We believed these first perceptions were biased by the views expressed by a few people who repeatedly contacted the South Dakota Department of Game, Fish and Parks to complain about the presence of lions (i.e., the squeaky wheel gets the grease); however, we had no information to support our assumption. The survey included about 1,100 responses from 1,783 residents surveyed (62% response rate). Responses were from South Dakota residents who lived in the Black Hills region as well as those from outside the region where mountain lions occurred (mostly eastern South Dakota). The responses to the survey indicated that there were generally two attitudes toward mountain lions: pro-lion and anti- or contra-lion (table 8.1). Over 50% of respondents to the survey were either slightly or strongly pro-lion. Many of these respondents equated the presence of the lion to a complete and healthy ecosystem for the Black Hills. Just over 30% were slightly or strongly anti-lion; about 4% were neutral or had no attitude toward the species. As might be expected, when we reviewed this finding, we wondered who made

Table 8.1.   Attitudes of South Dakotans toward mountain lions

Mountain lion attitude groups	Number	Percent
Strongly pro-lion	242	23%
Slightly pro-lion	360	34%
Neutral	120	11%
Slightly contra-lion	240	22%
Strongly contra-lion	105	10%

Source: Gigliotti, Fecske, and Jenks 2002.

up the two groups? Was there an age difference? Or maybe a regional difference; those from the Black Hills versus others? Or were these different attitudes due to life style: rural versus city?

Our first question was based on thoughts that younger individuals would be more pro-lion than older individuals because of the history of predators in South Dakota. In the past, lions and other large predators had been viewed as vermin and managed based on bounties (this was the case for lions in South Dakota prior to 1978). Thus, older South Dakota residents who were raised on farms and ranches had likely learned that mountain lions killed livestock and game species, and therefore those residents would be likely to have negative views toward lions. So we were not surprised to learn that the average age of respondents who were strongly contra-lion was about 10 years greater than for those who were strongly pro-lion (table 8.2).

We also learned that people who lived in rural areas and raised livestock tended to be strongly contra-lion, whereas those in other residential settings were more strongly pro-lion (table 8.3). The findings supported our suspicion that the potential for loss of livestock from mountain lions was an important driver of the contra-lion attitude. The

Table 8.2. *Average age of survey respondents classed by attitude toward mountain lions*

Mountain lion attitude groups	Mean Age	95% C.I.
Strongly pro-lion (23%)	50.3	48.7–52.0
Slightly pro-lion (34%)	49.3	47.8–50.9
Neutral (11%)	57.4	54.1–60.7
Slightly contra-lion (22%)	53.4	51.3–55.5
Strongly contra-lion (10%)	59.8	56.5–63.0
Group total	52.4	51.4–53.4

Source: Gigliotti, Fecske, and Jenks 2002.

Table 8.3. *Type of residence of respondents to a survey on attitudes toward mountain lions*

Type of residence	Mountain lion attitude group (group size)				
	Strongly pro-lion 23%	Slightly pro-lion 34%	Neutral 11%	Slightly contra-lion 22%	Strongly contra-lion 10%
City of more than 10,000 (40%)	37%	43%	45%	40%	28%
Town/city 2,000–10,000 (14%)	16%	14%	15%	11%	15%
Town less than 2,000 (15%)	13%	14%	24%	15%	16%
Suburban setting (10%)	12%	11%	6%	9%	8%
Rural, no livestock (12%)	15%	12%	7%	15%	6%
Rural with livestock (9%)	7%	6%	3%	10%	27%

Source: Gigliotti, Fecske, and Jenks 2002.

pro-lion attitude was a little stronger adjacent to or within lion country (counties in or near the Black Hills), compared to areas farther from lion country (counties east of the Missouri River) (table 8.4). This was probably because people living near lion country had a history of living near lions without experiencing any problems, while people farther away from lions could imagine potential problems from having lions living nearby.

The excitement connected with the presence of this new predator, as well as the fact that there had been limited negative interactions with the species, likely affected survey respondents' attitudes toward their preferred population status for mountain lions in the Black Hills. Those respondents who were pro-lion were much more likely to be comfortable with the current number of lions or would actually like to have more lions in the Black Hills (table 8.5). Respondents who held contra-lion attitudes preferred a slightly or greatly decreased population size of lions in the Black Hills.

South Dakota has a relatively high proportion of residents engaged with outdoor activities, including fishing (23%) and hunting (10%) (US Fish and Wildlife Service

Table 8.4. *Eastern South Dakota vs. western South Dakota (East River vs. West River) county residence analyzed by mountain lion attitude group*

	County residence	
Mountain lion attitude group (group size)	East River (70%)	West River (30%)
Strongly pro-lion (23%)	21%	28%
Slightly pro-lion (34%)	34%	34%
Neutral (11%)	12%	9%
Slightly contra-lion (23%)	23%	20%
Strongly contra-lion (10%)	10%	10%

Source: Gigliotti, Fecske, and Jenks 2002.

Table 8.5. *Preferred mountain lion population analyzed by mountain lion attitude group*

Mountain lion attitude group	Preferred mountain lion population					
	Decrease greatly	Decrease somewhat	Remain at current level	Increase somewhat	Increase greatly	No opinion
Strongly pro-lion	1%	0%	24%	53%	12%	11%
Slightly pro-lion	1%	4%	38%	33%	4%	21%
Neutral	3%	10%	59%	12%	0%	15%
Slightly contra-lion	4%	15%	38%	11%	1%	31%
Strongly contra-lion	43%	29%	16%	0%	2%	10%

Source: Gigliotti, Fecske, and Jenks 2002.

Table 8.6. *South Dakota residents' attitudes toward mountain lions,*
comparing big game hunters and nonhunters

	Normally hunt big game	
Mountain lion attitude groups	No (64%)	Yes (36%)
Strongly pro-lion (23%)	22%	24%
Slightly pro-lion (34%)	33%	35%
Neutral (11%)	12%	10%
Slightly contra-lion (22%)	24%	21%
Strongly contra-lion (10%)	9%	11%

Source: Gigliotti, Fecske, and Jenks 2002.

2011). Many residents hunt deer annually, and in the past, families, including residents and nonresidents, traditionally hunted deer in the Black Hills (Backman et al. 2001). Because this new predator could reduce deer and elk populations in the Black Hills or may even be perceived as a threat to personal safety when residents and their family members hunt in the Black Hills, we expected that big game hunters would have stronger negative attitudes toward mountain lions. However, when big game hunters were surveyed in 2002, we found that their attitudes toward mountain lions were similar to those of nonhunters (table 8.6). This finding could likely be explained by two facts: that mountain lion presence and sightings were relatively rare at the time and that deer hunting in the Black Hills was perceived as good in 2002 (Gigliotti 2003). The attitudes expressed in this survey made sense to me, based on the early conversations I had had with hunters. Some rather enjoyed just the thought that they might see a lion while out hunting for deer, elk, or another species. In contrast, some hunters would approach me, knowing I was studying the species, and speak about how they were unhappy to have lions occupying the Black Hills and, likely because they held a similar opinion to that of rural landowners with livestock, would like the species eliminated from the state. Overall, though, during these early years there was much more excitement associated with the presence of mountain lions than disdain for the predator.

The Later Years

In any research project focused on learning about aspects of a particular species, there is a scientific wonder that surrounds those actively involved in the study and others with whom they associate. This wonder can sometimes be misinterpreted as a concern for the species, a love of the species, or a need to protect it in some way. It seems that there is a threshold between two primary views of researchers who study large predators: they are seen either as unbiased researchers or as biased and exhibiting a protectionist mind-set. Unfortunately, no one knows for sure where the threshold separating the two views lies or just what one has to do to get bumped from unbiased researcher to protectionist. From our point of view, we might have classed our feelings, perceived by others as protectionist, more as worry about how future changes in management or the public's attitude toward mountain lions might affect current research projects. For example, what if someone (a child or a hunter) was killed by a lion? Would sentiment for the species change immediately from a tendency toward pro-lion attitudes to outright hatred for the species? If this happened, how would it affect our current projects? Would such a change result in a bounty on the species, subsequently forcing the end to current research projects because the lion population would be reduced to a level at which we could not maintain a reasonable sample of marked individuals?

At that time (i.e., 2010), some employees of the South Dakota Department of Game, Fish and Parks, as well as some hunters and South Dakota residents, seemed to think we were "protectors of the species" and thus more conservationist-minded or protectionist than just interested, focused scientists involved with the collection of information to inform the management agency (this was primarily where our interests lay). Even some Game and Fish Commission personnel intimated that I was against the harvest season and that my estimates of the size of the lion population were extremely conservative. This interpretation was totally incorrect; however, when I thought back on this categorization, I concluded that in fact I was being conservative in my estimates, but just as a result of my history with wildlife science and management. When obtaining my undergraduate and graduate degrees, I was taught that management should be conservative to ensure that populations were not overharvested. Thus, the accusation of conservative population estimates did fit with my views on management. Nevertheless, the new interpretations of our motives were not helpful when other factors within the Black Hills region changed and consequently affected the public's views on mountain lions.

Gigliotti (2009) evaluated the 2008 attitudes of Black Hills residents in a survey that included questions regarding the current status of the mountain lion population. Some of the questions were directly comparable to survey questions presented to South Dakota residents in 2002 (Gigliotti, Fecske, and Jenks 2002). For example, his new survey asked participants what size of mountain lion population they would prefer: did they want it increased or decreased, or were they comfortable with its present size? When Gigliotti (2012) compared responses from 2002 and 2008, he observed a shift from "remain the same" and/or "increase" to "decrease" or "remain the same" (table 8.7). This change in attitude toward the lion population occurred during the time period when the population was believed to be saturated (See chapter 6). Thus, the shift in public attitude was likely a response to increased numbers of news reports of sub-adult lions that were being removed by the South Dakota Department of Game, Fish and Parks biologists, additional reports of sightings of lions by the public (both residents and recreationalists), and news of lions being killed by vehicles on the extensive roadways that traverse the Black Hills region. Furthermore, the harvest limit had been met each year since the season had been established, further indicating that there were plenty of mountain lions in the Black Hills region.

Gigliotti (2012) compared the attitudes of South Dakota residents surveyed in 2002 (Gigliotti, Fecske, and Jenks 2002) with the attitudes of residents he surveyed in 2012. In many of the general attitudes toward mountain lions, there was little change. For example, the overall South Dakota residents' response to the question "Having a healthy, viable population of mountain lions in South Dakota is important to me" in 2002 was relatively similar to the response received in the 2012 survey (table 8.8). Also, responses to the question "Having any mountain lions in South Dakota is too dangerous a risk to people" (table 8.9) were similar in the two time periods (Gigliotti 2012). This finding indicated that most South Dakota residents did not consider mountain lions a threat, despite attacks documented in other states, such as California (Torres et al. 1996), and other regions (Beier 1991). One potential reason for these responses could be that there had been no verified attacks on humans by wild mountain lions (there had been one or more historical attacks by captive mountain lions) in the Black Hills; one suspected attack by a wild lion was unverified. When this one

Table 8.7. Comparison of attitudes of South Dakota residents in 2002 with Black Hills residents surveyed in 2008

Desired mountain lion population for the Black Hills	2002 South Dakota residents	2008 Black Hills residents
Decrease greatly	11%	14%
Decrease slightly	13%	30%
Remain the same	42%	46%
Increase slightly	28%	8%
Increase greatly	6%	2%

Source: Gigliotti 2012.

Table 8.8. Responses (2002 and 2012 surveys) to the statement "Having a healthy, viable population of mountain lions in South Dakota is important to me"

Attitude response	2002 SD residents	2012 SD residents
Strongly agree	13%	11%
Moderately agree	15%	16%
Slightly agree	19%	22%
Neutral/no opinion	28%	25%
Slightly disagree	7%	7%
Moderately disagree	6%	7%
Strongly disagree	12%	12%

Source: Gigliotti 2012.

Table 8.9. Responses (2002 and 2012 surveys) to the statement "Having any mountain lions in South Dakota is too dangerous a risk to people"

Attitude response	2002 SD residents	2012 SD residents
Strongly agree	7%	7%
Moderately agree	7%	7%
Slightly agree	11%	13%
Neutral/no opinion	13%	16%
Slightly disagree	18%	13%
Moderately disagree	20%	21%
Strongly disagree	24%	23%

Source: Gigliotti 2012.

potential attack occurred, I had been asked to comment on the evidence. A young lion had allegedly attacked a person who was ice fishing at Sheridan Lake, just west of Rapid City, but officials were unable to find fresh sign attributed to a lion during a two-day search after the attack (Watson 2008). At the time, the attack invoked much media attention, but the fact that there was limited evidence in support of the attack and that attacks were undocumented in the Black Hills might have helped to keep attitudes related to the risk of attack to a minimum, despite other information made available to the public, such as the number of lions, harvest information, and sightings by residents. Furthermore, most attacks by mountain lions had involved children, and few were fatal (Beier 1991). Nevertheless, I recall many a mother confronting me after a presentation on mountain lions, stating she was concerned about the safety of her children and asking for advice on how to keep them safe when living or recreating in the Black Hills.

Many South Dakota residents were, however, concerned about the impact of mountain lions on game species. My interaction with a former volunteer (mentioned earlier) was just an echo of comments by other hunters who regularly attended Game and Fish Commission meetings to express their views on the need to reduce the number of mountain lions in the Black Hills or eliminate them entirely. This view might have been considered the rumblings of a few hunters who were unable to obtain deer permits (after years of being successful), but the sentiment was quantified by Gigliotti (2012). His survey asked South Dakota residents to respond to the question "I am concerned about mountain lions killing too many game animals" (table 8.10). The responses of residents in 2012, compared to those in 2002, shifted from some disagreeing with the statement to more agreement. I was able to view this more negative sentiment when I was invited to a Game and Fish Commission meeting in 2014 to present results from our work on mountain lions to the commissioners. The meeting was highly attended by both those in favor of increasing the harvest limit on mountain lions and those individuals, some of whom were members of the Black Hills Mountain Lion Foundation, who voiced opposition to any increase in the harvest limit.

Table 8.10.   *Responses (2002 and 2012 surveys) to the statement "I am concerned*
about mountain lions killing too many game animals"

Attitude response	2002 SD residents	2012 SD residents
Strongly agree	6%	14%
Moderately agree	7%	12%
Slightly agree	12%	19%
Neutral/no opinion	24%	22%
Slightly disagree	15%	12%
Moderately disagree	16%	12%
Strongly disagree	20%	9%

Source: Gigliotti 2012.

Despite the information presented to the commission, the individuals who testified and were in favor of increasing the lion harvest consistently argued that because
the lion population had increased, these lions had reduced game populations, and that
these predators were killing a substantial number of harvestable game animals. For
example, their reasoning was that if there were 150 adult lions in the Black Hills, and
each adult killed 52 deer-sized ungulates per year, then the total kill would be about
7,800 ungulates per year. Considering that the total harvest of deer in the Black Hills
was in the neighborhood of 4,000 animals, it would be rather easy to conclude that lion
kills were affecting license tag allocation. The opposition testified that their review of
the scientific literature had provided no information supporting the statement that
large predators negatively affect game populations. However, just the potential for a
negative impact directly resulting from lions was compelling to the commissioners,
who voted to increase the harvest limit in that and subsequent years. It also resulted in
new research projects to assess the impact of lions on prey populations both through
the study of lions (Smith 2014) and through studies that documented cause-specific
mortality on elk in the Black Hills (e.g., Simpson 2015). Thus, the South Dakota Department of Game, Fish and Parks realized that there was a need for information specific to the Black Hills to support current and future management of mountain lions.

When attempting to quantify the total kill by mountain lions, it becomes easy to
ignore the potential for other mortality agents to kill animals that serve as prey to predators. Other mortality agents include disease, humans (vehicle mortalities and harvest), accidents, malnutrition, and severe weather, to name some of the common factors
(See chapter 4). Since mountain lions scavenge carcasses, including, for example, deer
that are killed by vehicles and those that die from diseases (such as chronic wasting
disease), a certain number of prey consumed by lions would be expected to have died
from other causes and then to have been scavenged by lions during the year; and, as
presented earlier, the scavenging rate for lions in the Black Hills was significantly

higher than that documented for other mountain lion populations under study. Even before mountain lions were killing deer in the central Black Hills, DePerno et al. (2000) estimated the annual survival of adult female white-tailed deer at about 57%. The harvest rate was estimated at 10%, which meant that 33% of adult females were dying from other causes; in that study, those causes were coyote and domestic dog predation, malnutrition, sickness, and other unidentified agents of mortality. If there were 40,000 deer in the Black Hills, then about 13,000 deer would have been dying from causes other than harvest before the time when lions became a dominant predator in the region. As a result, the predatory impact of lions on prey in the Black Hills region could be negligible, because other sources of mortality would be compensating for predation.

The objective of one of our studies (Simpson 2015) was directly related to assessing the effects of mountain lions on elk in the Black Hills. The individuals who testified that mountain lions were reducing game populations had used this big game species as an example of the impact of lions. The elk population had been reduced in the recent past by a high level of harvest because of complaints from landowners who were experiencing damage (loss of crops, damage to fences) to their ranches. However, the reduced number of elk also was associated with lions, possibly because of a decreased rate of growth, otherwise known as a "predator trap" (e.g., Gasaway et al. 1992) perpetuated by a relatively high lion population. Some evidence in support of this hypothesis had been collected by South Dakota Game, Fish and Parks biologists who were conducting a study on elk in Custer State Park. Their initial findings indicated low survival of elk calves owing to lion predation. We were funded to conduct a companion project on elk west of Jewel Cave National Monument and north of Highway 16 to the Deerfield Lake Region of the Black Hills (Simpson 2015). The study area represented over 30% of the Black Hills region in South Dakota, and surveys indicated that there were as many as 1,000 elk occupying the area. We radio-collared 40 cow elk per year for two years and subsequently collared close to 40 elk calves per year for the study. Results of our study indicated that elk calf survival averaged 75% and was actually higher than had been found in most elk populations in western states; it was equivalent to what had been found for elk calf survival in eastern states (table 8.11). Furthermore, the average weight of cow elk in this region of the Black Hills was high, another indication that this population was doing well. Thus, the conclusion from the study was that elk were not being affected to a significant degree by mountain lions. Nevertheless, of the mortality that was documented on these calves, at least 75% was from predation; thus, lions were killing some of the calves. It should be noted that the lion population was in the process of being reduced while we were conducting our study, and thus, any reduction in the lion population could have resulted in fewer elk calves killed by lions. Furthermore, in contrast to what had been presented at the Game and Fish Commission meeting, there was some published information that indicated that reducing

Table 8.11. *Comparison of elk calf survival rates among elk populations throughout North America*

Area	Summer	Winter	Annual	Source
Eastern populations				
Kentucky			0.77	Seward 2003
Michigan	0.90	0.97	0.87	Bender et al. 2002
Pennsylvania	0.92	0.90	0.82	DeVivo et al. 2011
Pennsylvania			0.71	Cogan 1999
North Carolina			0.59	Murrow, Clark, and Delozier 2009
Western populations				
California	0.85			Howell et al. 2002
North-central Idaho	0.18–1.00			White, Zager, and Gratson 2010
North-central Idaho	0.00–0.84		0.06–0.83	Zager, White, and Pauley 2005
North-central Idaho	0.32			Schlegel 1976
Montana		0.82–0.86		Knight 1970
Northern Yellowstone	0.65	0.72	0.43	Singer et al. 1997
Northern Yellowstone	0.29	0.90	0.22	Barber-Meyer, Mech, and White 2008
Northwestern Wyoming	0.84	0.84	0.58	Smith and Anderson 1998
Northwestern Wyoming		0.26–0.69		Sauer and Boyce 1983
Southeastern Washington			0.47	Myers 1999
South Dakota	0.79	0.96	0.75	Simpson 2015

Source: Simpson 2015.

numbers of mountain lions reduced the risk of mortality to elk calves (White, Zager, and Gratson 2010).

From the start of our studies on mountain lions, we always expected that our results would be directly linked to management of the species and that a harvest season would be enacted to manage the species to a desired level by providing hunting opportunities to the residents of the state. That view was echoed by residents who were surveyed in 2012 (Gigliotti 2012). Although there had always been relatively strong support for a harvest season on mountain lions, there was a substantial amount of opposition to a harvest when it was initially proposed. That opposition resulted in a suit to obtain an injunction to stop the first harvest in 2005. However, the suit was overturned, and harvest seasons continued unobstructed after that year. By 2012, any suspicion that the harvest would result in extinction of the species (see chapter 4), inbreeding owing to reduced population size (see chapter 7), or harmful diseases that would negatively affect the population (see chapter 5) seemed to have disappeared from the public's view. In fact, when residents were again asked the question "Do you oppose or favor a regulated mountain lion season in South Dakota?," the same percent-

Table 8.12. Responses (2002 and 2012 surveys) to the question "Do you oppose or favor a regulated mountain lion season in South Dakota?"

Attitude response	2002 SD residents	2012 SD residents
Strongly oppose	7%	5%
Moderately oppose	4%	1%
Slightly oppose	3%	2%
Neutral/no opinion	14%	20%
Slightly favor	15%	19%
Moderately favor	26%	19%
Strongly favor	31%	34%

Source: Gigliotti 2012.

age of South Dakota residents (72%) who in 2002 favored a regulated mountain lion season for managing mountain lions in South Dakota, did so again in 2012 (table 8.12). I believe that favorable view could be attributed to the information collected on the species, which was regularly shared with the public through the South Dakota Game, Fish and Parks public forums and our presentations to local organizations across the state. The result was an informed public who were comfortable with the methods used by the South Dakota Department of Game, Fish and Parks to manage what was considered a politically and physically dangerous animal while it reestablished its presence in the Black Hills.

Literature Cited

Backman, P. A., D. E. Hubbard, J. A. Jenks, and L. M. Gigliotti. 2001. The importance of the social aspects of hunting: A comparison between resident and non-resident 1999 Black Hills deer hunters. Proceedings of the South Dakota Academy of Science 80:267–281.

Barber-Meyer, S. M., L. D. Mech, and P. J. White. 2008. Elk calf survival and mortality following wolf restoration to Yellowstone National Park. Wildlife Monographs 169:1–30.

Beier, P. 1991. Cougar attacks on humans in the United States and Canada. Wildlife Society Bulletin 19:403–412.

Bender, L. C., E. Carlson, S. M. Schmitt, and J. B. Haufler. 2002. Production and survival of elk (*Cervus elaphus*) calves in Michigan. American Midland Naturalist 148:163–171.

Cogan, R. D. 1999. Elk population survey. Annual Report. Bureau of Wildlife Management, Pennsylvania Game Commission. Harrisburg, PA.

DelGuidice, G. D. 1998. Surplus killing of white-tailed deer by wolves in northcentral Minnesota. Journal of Mammalogy 79:227–235.

DePerno, C. S., J. A. Jenks, S. L. Griffin, and L. A. Rice. 2000. Female survival rates in a declining white-tailed deer population. Wildlife Society Bulletin 28:1030–1037.

DeVivo, M. T., W. O. Cottrell, J. M. DeBerti, J. E. Duchamp, L. M. Heffernan, J. D. Kougher, and J. L. Larkin. 2011. Survival and cause-specific mortality of elk *Cervus canadensis* calves in a predator rich environment. Wildlife Biology 17:156–165.

Gasaway, W. C., R. D. Boetje, D. V. Grandgaard, D. G. Kelleyhouse, R. O. Stephenson, and D. G. Larsen. 1992. The role of predation in limiting moose at low densities in Alaska and Yukon and implications for conservation. Wildlife Monographs 120:3–59.

Gigliotti, L. M. 2003. 2002 Black Hills deer hunter survey. HD-4-03.AMS. Pierre, SD: Game, Fish and Parks Department, Division of Wildlife.

Gigliotti, L. M. 2009. 2008 Black Hills deer hunter survey. HD-2-09.AMS. Pierre, SD: Game, Fish and Parks Department, Division of Wildlife.

Gigliotti, L. M. 2012. Public perceptions of mountain lions and their management in South Dakota (2002–2012). Unpublished report provided to South Dakota Department of Game, Fish and Parks. South Dakota Cooperative Fish and Wildlife Research Unit, Brookings, SD.

Gigliotti, L. M., D. M. Fecske, and J. A. Jenks. 2002. Mountain lions in South Dakota: A public opinion survey—2002. HD-9-02.AMS. Pierre, SD: South Dakota Department of Game, Fish and Parks.

Howell, J. A., G. C. Brooks, M. Semenoff-Irving, and C. Greene. 2002. Population dynamics of Tule elk at Point Reyes National Seashore, California. Journal of Wildlife Management 66:478–490.

Knight, R. R. 1970. The Sun River elk herd. Wildlife Monographs 23:3–66.

Kruuk, H. 1972. Surplus killing by carnivores. Journal of Zoology 166:233–244.

Murrow, J. L., J. D. Clark, and E. K. Delozier. 2009. Demographics of an experimentally released population of elk in Great Smoky Mountains National Park. Journal of Wildlife Management 73:1261–1268.

Myers, W. D. 1999. An assessment of elk population trends and habitat use with special reference to agricultural damage zones in the northern Blue Mountains of Washington. Final Report. Olympia, WA: Washington Department of Fish and Wildlife.

Sauer, J. R., and M. S. Boyce. 1983. Density dependence and survival of elk in northwestern Wyoming. Journal of Wildlife Management 47:31–37.

Schlegel, M. 1976. Factors affecting calf elk survival in north central Idaho: A progress report. Proceedings of the Western Association of State Game Fish Commission 56:342–355.

Seward, N. W. 2003. Elk calf survival, mortality, and neonatal habitat use in eastern Kentucky. Master's thesis, University of Kentucky.

Simpson, B. D. 2015. Population ecology of Rocky Mountain elk in the Black Hills, South Dakota and Wyoming. Master's thesis, SDSU.

Singer, F. J., A. Harting, K. K. Symonds, and M. B. Coughenour. 1997. Density dependence, compensation, and environmental effects on elk calf mortality in Yellowstone National Park. Journal of Wildlife Management 61:12–25.

Smith, B. L., and S. H. Anderson. 1998. Juvenile survival and population regulation of the Jackson elk herd. Journal of Wildlife Management 62:1036–1045.

Smith, J. B. 2014. Determining impacts of mountain lions on bighorn sheep and other prey sources in the Black Hills. PhD diss., SDSU.

Torres, S. G., T. M. Mansfield, J. E. Foley, T. Lupo, and A. Brinkhaus. 1996. Mountain lion and human activity in California: Testing speculations. Wildlife Society Bulletin 24:451–460.

US Fish and Wildlife Service. 2011. 2011 National Survey of Fishing, Hunting, and Wildlife-Associated Recreation. https://www.census.gov/prod/2012pubs/fhw11-nat.pdf.

Watson, M. 2008. Rapid City man attacked by mountain lion. Spearfish (SD) Black Hills Pioneer, March 3.

White, C. G., P. Zager, and M. W. Gratson. 2010. Influence of predator harvest, biological factors, and landscape on elk calf survival in Idaho. Journal of Wildlife Management 74:355–369.

Zager, P., C. White, and G. Pauley. 2005. Elk ecology: July 1, 2005 to June 30, 2006. Study IV, Factors influencing elk calf recruitment. Boise: Idaho Department of Fish and Game.

Epilogue

As the human population on earth approaches 8 billion individuals, dispersed within just about every crevasse that can be occupied, it is refreshing to know that a large predator, such as the mountain lion, can still make its way to and subsequently recolonize a region that was once a central part of its distribution. Considering the dispersal lengths documented in this (Thompson and Jenks 2005, 2010; Hawley et al. 2016) and other studies (Stoner et al. 2010), if the mountain lion was provided the opportunity to recolonize other regions, no human intervention would be required to ensure success. Nevertheless, the need to traverse inhospitable habitat punctuated with high-speed highways and ever-expanding cities and towns will likely make future journeys by dispersing male and female mountain lions difficult and rarely successful.

It is not known whether mountain lions consistently occupied the Black Hills or whether they were extirpated at some time and later dispersers from Wyoming recolonized the region. We do know, based on the work accomplished on genetics, that the population was founded by just a few individuals (Thompson 2009; Juarez et al. 2016). Those first individuals were rarely seen, and some were likely killed, possibly because they were mistaken for game animals or because the presence of the species was considered a threat to livestock (fig. E.1). One such individual was an adult female with kittens that was found dead early in our work in the southern Black Hills. Actions such as this may have kept the population low from 1978, when the bounty was removed from the species, until the late 1990s, when the population began to increase and we began our study of the species. At that time, few people, either locals within the Black Hills or employees of South Dakota Game, Fish and Parks, believed there were many lions in the area or that the species had the capability to increase to the extent that has been documented.

FIGURE E.1. Ponderosa pine trees provide ubiquitous sites that are used by mountain lions to hide from recreationalists in the Black Hills. *Photo by Emily Mitchell.*

Once the recolonization of the Black Hills began, the lion population exhibited characteristics that have been documented for other species, such as white-tailed deer, which in the Black Hills came to represent the primary prey of this predator. Other studies have documented the large litter size (up to six kittens [Anderson 1983]) and a high rate of increase (Logan and Sweanor 2000) for the species, so it should not have been surprising that, given sufficient prey, population expansion could be rapid. Therefore, once enough individuals colonize an area, such as the Black Hills, little time elapses before a population of sufficient size develops. At low population size, males seem to traverse large expanses to seek out and mate with females, which seem to be poorly defended, based on the few battle scars documented on the first males captured during our studies. Kittens have relatively high survival, possibly because of the diverse prey resources that occur in the Black Hills region. During initial establishment, lions can depend on easy prey resources such as porcupines (Fecske, Jenks, and Lind-zey 2003) and those they are skilled at killing, such as mule deer. Once established and forced to adapt because of higher population size and depleted focal prey species, mountain lions can and do learn to kill other prey, such as white-tailed deer, bighorn sheep, mountain goats, and elk. Because these prey species were not distributed uniformly throughout the Black Hills, some lions became adept at killing common prey in their subregion or home area (Smith 2014).

Population characteristics such as growth rate, primary prey, reproductive rate, home range size, disease prevalence, genetic diversity, and nutritional condition were evaluated during this time period, which began in the 1990s and extended to 2015. Most of the mortality documented during our study, even prior to the initiation of a harvest, was human-related (Thompson, Jenks, and Fecske 2014). Even so, the limited harvest initiated on the species had little or no impact on population size and genetic diversity but likely improved nutritional condition and reduced disease prevalence within the population. Population improvement could have been the result of increased prey availability caused by reduced population size. The mortality caused by both agents reduced the dependency of mountain lions on domestic prey and potentially increased the availability of free-ranging prey, thus limiting exposure of mountain lions to diseases prevalent in pets, such as domestic cats and dogs. Consequently, our studies illustrated how, in regions with multiple species of prey, a mountain lion population can expand at a high rate over a short period of time and, when necessary, adapt to changing conditions.

Later in our investigations, the harvest rate on mountain lions was increased. Based on various indices (including nutritional condition and disease prevalence) and population modeling, the increased harvest likely resulted in a decline in population size. The expected outcome of harvest was reduced genetic diversity, an increased rate of genetic deformities, and population extinction. None of these effects, however, were documented. To the contrary, we documented that genetic diversity was stable and relatively high when compared to other mountain lion populations. We suspect that genetic diversity was conserved because of adequate movement of mountain lions within and into the Black Hills region, likely from the north and the west. As a result, we recognize that the conservation of genetic diversity and population quality can be affected by factors beyond the control of local wildlife professionals.

The relationship between some humans and mountain lions is unique because of the resemblance of young lions to domestic cats. Because of this similarity, I encountered a number of people who asked me about the feasibility of raising mountain lions in their homes. I always responded to the question with a "no" and asked them to think about the 100–150-pound adult cat that would be lying on their sofa in just over a year rather than the cute 5-pound kitten they would be bringing into their household. For example, a captive male mountain lion I acquired at 6 months of age (weight was about 16 kg [39 lbs.]) that had been raised in a household in Montana would consistently move away when I approached his pen but often would hiss and attack when I was feeding him if our eyes met when we were near each other. In contrast, his sibling (18 kg [44 lbs.] at 6 months of age) would just attempt to hide from me despite weighing substantially more than his hostile brother (both weighed more than 42 kg [100 lbs.] when they were transferred to a zoo). These interactions, along with those of a number of kittens I raised to 6 months of age or older, convinced me that young as well as adult mountain lions are unpredictable, respond innately to sounds and actions, and despite a desire to move away when in threatening situations, can attack even the most devoted owner who has no intention to cause harm to the animal.

Because of their size, secrecy, and adaptations for killing and consuming large prey, mountain lions elicit positive and negative responses from people. When mountain lions were initially recolonizing the Black Hills, some of the residents of the Black Hills and South Dakota appeared excited about the presence of this new species. Many considered the Black Hills to be a functioning ecosystem that was replete with prey and their natural predators. However, this feeling of awe changed somewhat as the size of the mountain lion population increased and the potential effects of mountain lions

FIGURE E.2. Mountain lions generally loaf (sleep) during the day and become active at night. *Photo by Dan Thompson.*

on reduced populations of deer, elk, and other prey species were circumstantially linked. Our studies barely touched upon these interactions. Nevertheless, the complexities of systems with predators and their prey require an understanding of all the components and their interactions (fig. E.2).

Studying mountain lions for over a decade allowed my students and me to document how the species could expand from just a few individuals in the mid-1990s to what we termed a saturated population in a relatively short period of time. We witnessed how the species learned to capture and consume new prey species when prey availability changed, and how it could remain viable when the population was reduced through harvest. There will always be a mystique associated with a species that can weigh up to 175 lbs., that travels about its home area mostly at night, and that is secretive in its nature. Our work has provided insight into how it successfully recolonized a small corner of its distribution in North America, the Black Hills. I have confidence that with continued management the species will thrive and provide future generations the thrill of seeing this charismatic critter in its natural environment.

Literature Cited

Anderson, A. E. 1983. A critical review of literature on puma (*Felis concolor*). Special Report 54, Colorado Division of Wildlife, Littleton, CO.

Fecske, D. M., J. A. Jenks, and F. G. Lindzey. 2003. Characteristics of mountain lion mortalities in the Black Hills, South Dakota. In Proceedings of the Sixth Mountain Lion Workshop, 25–29. Austin: Texas Parks and Wildlife Department.

Hawley, J. E., P. W. Rego, A. P. Wydeven, M. K. Schwartz, T. C. Viner, R. Kays, K. L. Pilgrim, and J. A. Jenks. 2016. Long-distance dispersal of a subadult male cougar from South Dakota to Connecticut documented with DNA evidence. Journal of Mammalogy 96:1435–1440.

Juarez, R. L., M. K. Schwartz, K. L. Pilgrim, D. J. Thompson, S. A. Tucker, and J. A. Jenks. 2016. Assessing temporal genetic variation in a cougar population: Influences of harvest and neighboring populations. Conservation Genetics 17:379–388.

Logan, K. A., and L. L. Sweanor. 2000. Puma. In ecology and management of large mammals in North America, ed. S. Demarais and P. R. Krausman, 347–377. Upper Saddle River, NJ: Prentice Hall.

Smith, J. B. 2014. Determining impacts of mountain lions on bighorn sheep and other prey sources in the Black Hills. PhD diss., SDSU.

Stoner, D. C., W. R. Rieth, M. L. Wolfe, M. B. Mecham, and A. Neville. 2010. Long-distance dispersal of a female cougar in a basin and range landscape. Journal of Wildlife Management 72:933–939.

Thompson, D. J. 2009. Population demographics of cougars in the Black Hills: Survival, dispersal, morphometry, genetic structure, and associated interactions with density dependence. PhD diss., SDSU.

Thompson, D. J., and J. A. Jenks. 2005. Long-distance dispersal by a subadult male cougar (*Puma concolor*) from the Black Hills, South Dakota. Journal of Wildlife Management 69:818–820.

Thompson, D. J., and J. A. Jenks. 2010. Dispersal movements of subadult cougars from the Black Hills: The notions of range expansion and recolonization. Ecosphere 1:1–11.

Thompson, D. J., J. A. Jenks, and D. M. Fecske. 2014. Prevalence of human-caused mortality in an unhunted cougar population and potential impacts to management. Wildlife Society Bulletin 38:341–347.

INDEX

Page locators in italics signify photos, figures, or tables.